Tes

Watson, Billingsley, Croft, and Huntsberger

STATISTICS FOR MANAGEMENT AND ECONOMICS

Fifth Edition

prepared by

Clark Kristensen
Indiana University and Purdue University at Indianapolis

Allyn and Bacon
Boston • London • Toronto • Sydney • Tokyo • Singapore

Contents

Preface

This *Test Bank* is designed to accompany the fifth edition of <u>Statistics for Management and Economics</u> by Collin J. Watson, Patrick Billingsley, D. James Croft, and David V. Huntsberger. It contains over 1,000 multiple-choice and open-ended questions and has been updated thoroughly to reflect changes in the fifth edition. Test bank questions are identified by chapter number and level of difficulty and appear in the sequence in which the material appears in the textbook. For each section of each chapter, an adequate mix of questions by level of difficulty (Easy, Medium, and Challenging) is provided so that even the most highly able and diligent students should be challenged by the more difficult questions.

Open-ended questions are included in the test bank to give instructors greater flexibility in testing for higher learning objectives that is often difficult to do using the conventional multiple-choice question framework. I believe that such questions are an important part of any test bank for a course in business statistics. Several concepts or tools that are essential to a good understanding of statistics (the central limit theorem, confidence intervals, hypothesis tests, and regression analysis to name a few), are conducive to open-ended questions, making testing for higher learning objectives both necessary and desirable.

Questions in the *Test Bank* are available with test-generating software so that tests can be created quickly either by selecting questions by random selection or by chapter and question number. The flexibility of the *Test Bank* allows instructors to combine questions from multiple chapters to form a single test, scramble questions to create different versions of the same test, or edit questions to better suit their needs and desires.

The contribution of Hazem Arafa for preparing the fourth edition of the test bank is greatly appreciated. In addition, the comments I received from reviewer Gwynn Evans and Professor Watson were of great benefit to me in identifying and correcting mistakes in earlier versions of the test bank. These comments have been incorporated into the fifth edition of the test bank. Special thanks go to editor Rich Wohl for allowing me to do this project and to Dominique Vachon for all of her help and support throughout the duration of the project.

Clark E. Kristensen

Introduction to Statistics

1.1.1: Statistical science

(a) can only be applied to the field of economics
(b) is unimportant for anyone going into business
(c) is often used to aid in business decision making
(d) is not used by especially keen people in business
(e) primarily deals with the collection, analysis, and
     interpretation of population data

Answer: (c)
Difficulty level: Easy

1.1.2: Which of the following represents an application of statistical methods to aid in business decision making?

(a) managers using methods of statistical quality control
     to manage and constantly improve production processes
(b) accountants using statistical methods with financial
     ratio data to analyze the financial condition of
     corporations
(c) economists using statistical methods to develop indices
     to measure inflation and unemployment over time
(d) manufacturers conducting marketing research to collect
     and analyze data relating to the marketing of goods and
     services
(e) all of these are applications of statistical methods to
     aid in business decision making

Answer: (e)
Difficulty level: Easy

1.1.3: When there is uncertainty as to the conclusions that should be made during business decision making, _____ is (are) used to evaluate the reliability of conclusions based on data.

(a) descriptive statistics
(b) concepts of probability
(c) exploratory data analysis
(d) special topics in statistics
(e) regression analysis

Answer: (b)
Difficulty level: Medium

1.1.4: Which of the following provides the best definition of statistics?

(a) Statistics deals with using descriptive statistics and probability theory to aid in business decision making under certainty.
(b) Statistics is the science that deals only with parametric tests.
(c) Statistics is the science that deals with collecting, analyzing, and interpreting data.
(d) Statistics is the science that deals with making inferences about sample statistics using population parameters.
(e) None of these.

Answer: (c)
Difficulty level: Easy

1.1.5: Which of the following represent main areas of statistical methods?

(a) statistical inference
(b) probability
(c) descriptive statistics
(d) statistical inference and descriptive statistics only
(e) all of these

Answer: (e)
Difficulty level: Easy

1.2.1: *Descriptive Statistics* deals with:

(a) methods of organizing data only
(b) methods of presenting data only
(c) methods of summarizing data only
(d) methods of organizing, presenting, and summarizing data
(e) methods that provide a foundation for probability theory

Answer: (d)
Difficulty level: Easy

1.2.2: Which of the following statements below are correct?

I.   A *parameter* is an overall summary measure applied to a
      population whereas a *statistic* designates a summary
      measure applied to a sample

II.   A census is a sampling procedure often used because of
      its comprehensiveness and low cost

III. Managers who are knowledgeable about risk and
      variability, and who think effectively under conditions
      of uncertainty, can improve the profitability and
      performance of their firms

(a)   I only
(b)   II only
(c)   III only
(d)   I and III only
(e)   II and III only

Answer: (d)
Difficulty level: Easy to Medium

1.2.3: A manager is asked to perform the tasks outlined
below. Indicate the type of statistical problem the manager
is solving: descriptive statistics, probability, statistical
inference, or statistical technique.

a.   Condense a set of three hundred personal income figures
      into a meaningful set of charts and graphs.

b.   Estimate the size of the average overdraft for
      noncommercial customers at a bank based on a sample of
      overdrafts taken during the past three months.

c.   Find the proportion of defects being produced on an
      assembly line by taking a random sample of output.

d.   Predict the effect of an increase in a company's
      advertising budget on its annual gross sales.

e.   Measure or determine seasonal differences in a
      company's ice cream sales for the period 1986-1991.

Answer: a.   Descriptive Statistics
           b.   Inference
           c.   Inference
           d.   Statistical Techniques
           e.   Statistical Techniques
Difficulty level: Medium

1.3.1: The main reason for using random surveys is:

(a)   to directly control the selection process
(b)   to elicit true answers to sensitive questions
(c)   to reduce self-selection bias and nonresponse bias
(d)   to make sure that everyone in the population is sampled
(e)   all of these

Answer: (b)
Difficulty level: Easy

1.3.2: *Quantitative* data

(a)   are ordinal data that classify items into categories
      that can be ranked
(b)   are nominal data that classify items into one of a set
      of categories
(c)   are ratio data measured on a ratio-preserving scale
(d)   are interval data that have an equal and fixed distance
      between points on the measurement scale
(e)   can be subdivided into interval or ratio data

Answer: (e)
Difficulty level: Easy

1.3.3: *Qualitative* data

(a)   are ordinal data that classify items into categories
      that can be ranked
(b)   are nominal data that classify items into one of a set
      of categories
(c)   are ratio data measured on a ratio-preserving scale
(d)   are interval data that have an arbitrary zero point
(e)   can be subdivided into nominal or ordinal data

Answer: (e)
Difficulty level: Easy

1.3.4: Which of the following best describes a basic DIFFERENCE between nominal data and ordinal data?

(a)   Nominal and ordinal data are both quantitative data
(b)   Nominal and ordinal data are both qualitative data
(c)   Nominal data are qualitative data whereas ordinal data are quantitative data
(d)   Nominal data have a meaningful zero point and the ratio of the two data values is meaningful whereas with ordinal data the zero point is arbitrary
(e)   Nominal data classify items into one of a set of categories whereas ordinal data classify items into categories that can be ranked or ordered

Answer: (e)
Difficulty level: Easy

1.3.5: Which of the following best describes something that interval data and ratio data have in common?

(a)   Interval and ratio data are both quantitative data.
(b)   Interval and ratio data are both qualitative data.
(c)   Interval data are qualitative data. Ratio data are quantitative data.
(d)   Ratio data have a meaningful zero point and the ratio of the two data values is meaningful whereas interval data have an equal and fixed distance between points on the measurement scale but the zero point is arbitrary.
(e)   Ratio data classify items into one of a set of categories whereas interval data classify items into categories that can be ranked or ordered.

Answer: (a)
Difficulty level: Easy

1.3.6: Which of the following best describes a basic DIFFERENCE between interval data and ratio data?

(a)   Interval and ratio data are both quantitative data.
(b)   Interval and ratio data are both qualitative data.
(c)   Interval data are qualitative data. Ratio data are quantitative data.
(d)   Ratio data have a meaningful zero point whereas with interval data the zero point is chosen arbitrarily.
(e)   None of these.

Answer: (d)
Difficulty level: Medium

1.3.7: Ann Landers once asked her readers "If you had to do it over again, would you have children?" She received 10,000 responses, 70% of which said no they would not. In interpreting this sample information as an indication of parents' feelings about children, which of the following statistical concepts should be kept in mind?

(a)   non-response bias
(b)   self-selection bias
(c)   hidden differences
(d)   induced bias
(e)   all of these

Answer: (b)
Difficulty level: Medium

1.3.8: The study where the investigator examines variables of interest by using observed or historical data is called:

(a)   a quantitative study
(b)   an observational study
(c)   an experimental study
(d)   a qualitative study
(e)   a categorical study

Answer: (b)
Difficulty level: Easy

1.3.9: The type of sample that is easiest to analyze and widely used because of its' desirable characteristics is:

(a)   a census
(b)   a stratified sample
(c)   a convenience sample
(d)   a cluster sample
(e)   a simple random sample

Answer: (e)
Difficulty level: Easy

1.3.10: For each of the following, identify the type of data and whether it is discrete or continuous:

a.   Gender of all United States citizens.
b.   Educational attainment of New York residents (no college degree, undergraduate college degree, graduate college degree).
c.   Combat readiness of infantry soldiers in the United States Army, measured on a 0-100 scale.
d.   Exact temperature at 1:00pm on August 15, 1992 for ten major cities in the United States, in degrees celsius.
e.   Annual expenditures for the United States government, by source, to the nearest dollar.
f.   Exact distance between New York City and Los Angeles.
g.   Weekly figures for the Consumer Price Index (CPI), the Producer Price Index (PPI), and the national rate of unemployment, measured to the nearest tenth of 1%.

Answer: a.   discrete, nominal
           b.   discrete, ordinal
           c.   discrete or continuous, interval
           d.   continuous, interval
           e.   discrete, nominal
           f.   continuous, ratio
           g.   discrete, nominal
Difficulty level: Easy to Medium

1.3.11: In which publications are the following pieces of information most likely to be found?

Statistical Abstract of the United States (SA)
Survey of Current Business (SCB)
Encyclopedia of Associations (EA)

_____   a.   The combined sales volume of all retailers in the United States during December 1991.
_____   b.   Electrical power generated in each state in the United States over the past five years.
_____   c.   Heating oil inventory in the United States by month for the winter months of last year.
_____   d.   The list of publications put out by the American Management Association.

Answer: a.   SCB
           b.   SA
           c.   SCB
           d.   EA
Difficulty level: Easy

1.3.12: Which of the following best describes something that nominal data and ordinal data have in common?

(a)   Nominal and ordinal data are both quantitative data.
(b)   Nominal and ordinal data are both qualitative data.
(c)   Nominal data are qualitative data whereas ordinal data are quantitative data.
(d)   Nominal data have a meaningful zero point whereas with ordinal data the zero point is chosen arbitrarily.
(e)   Nominal data classify items into one of a set of categories whereas ordinal data classify items into categories that can be ranked or ordered.

Answer: (b)
Difficulty level: Easy

1.3.13: A survey that attempts to include every element of the population is called:

(a)   a census
(b)   a stratified sample
(c)   a convenience sample
(d)   a cluster sample
(e)   a simple random sample

Answer: (a)
Difficulty level: Easy

1.3.14: The study where the investigator *directly* controls or determines which subjects or experimental units or material receive treatments that are thought to affect variables of interest is called a(n):

(a)   quantitative study
(b)   observational study
(c)   experimental study
(d)   qualitative study
(e)   categorical study

Answer: (c)
Difficulty level: Easy

1.4.1: For each of the following, identify the type of data
and then recommend appropriate statistical graphical methods
for presenting this data:

a.   Standard and Poors 500 index, weekly figures for 1991
b.   Annual budget expenditures for the United States
      government, by source, to the nearest dollar, 1990-91
c.   Monthly number of small businesses filing for
      bankruptcy in 1991
d.   Actual versus projected annual sales for a small
      pharmaceutical company, 1987-1991
e.   Annual sales for the past two years of operation for a
      small and growing consulting company

Answer: a.   discrete nominal, bar chart
           b.   continuous nominal, pie and bar charts
           c.   discrete nominal, bar chart
           d.   continuous nominal data, grouped bar chart
           e.   continuous nominal data, pictorial chart
Difficulty level: Easy to Medium

1.4.2: The Board of regents at a liberal arts college have
become concerned with declining enrollments during recent
years. You are hired as a statistician and told to submit a
report containing a brief graphical presentation on annual
enrollment patterns at the college during 1985-1990. The
graphical method *best* suited for this purpose is the:

(a)   pie chart
(b)   bar chart
(c)   pictorial chart
(d)   grouped bar chart
(e)   histogram

Answer: (b)
Difficulty level: Easy to Medium

1.4.3: A(n) _____ chart can be used effectively to depict the
proportions or percentages of a total quantity corresponding
to several different categories. A(n) _____ chart is often
used to compare data values for two categories over time.

(a)   grouped bar; line
(b)   grouped bar; pie
(c)   pictorial; pie
(d)   line; pie
(e)   bar; pie

Answer: (b)
Difficulty level: Easy

1.6.1: The day after John Lennon, the leader of the famous "Beatles" rock group, was shot to death, a survey of 100,000 people showed that 90% of the public favored gun control. A gun manufacturer reviewing these results argued that the survey was not reliable. He was referring to the fact that:

(a)   a larger sample size was needed.
(b)   the wording of the questionnaire caused more people to favor gun control than would normally do so.
(c)   the response rate was too small to make inferences about the entire population.
(d)   if the survey were conducted at another time, different conclusions might result.
(e)   the survey did not measure what it was supposed to.

Answer: (d)
Difficulty level: Easy

1.6.2: A quality control manager inspects a large shipment of electrical fuses received from a new supplier. On the basis of the sample, the quality control manager accepts the shipment of electrical fuses. In arriving at this decision:

(a)   the quality control manager interpreted the fuses tested and the shipment both as samples
(b)   the quality control manager interpreted the fuses tested and the shipment both as populations
(c)   the quality control manager interpreted the fuses tested as a population and the shipment as a sample
(d)   the quality control manager interpreted the fuses tested as a sample and the shipment as a population
(e)   the quality control manager followed a procedure that the concepts of population and sample do not apply

Answer: (d)
Difficulty level: Medium

1.6.3: Consider the following list of statistical mistakes:

a.   Failure to adjust to a per-item basis (PIB)
b.   Forgetting to adjust for inflation (INF)
c.   Induced bias (IB)
d.   Self-selection (SS)
e.   Hidden differences (HD)

In the situations below, indicate the type of mistake that was made. If no apparent mistake has been made, write "NM."

_____ (1)   Forty-five women were randomly selected to test a new diet. They were offered the diet and monitoring of their progress free of charge. Thirty women accepted the offer and lost an average of 7.2 pounds per woman.

_____ (2)   A survey of people showed that those who had never been married were healthier than those who were married or who had been married at some time in the past. Thus it appears that marriage is bad for your health.

_____ (3)   Harriet Pushi found that when she surveyed 125 men for their attitudes on the Equal Rights Amendment, 100 agreed with her that it should be passed.

_____ (4)   One-hundred thirty mothers and their preschool children were tested to determine their levels of self-esteem in order to determine if a relationship existed between the two groups. The conclusion was that no significant relationship existed, that generally the children scored much higher than the mothers. However, a closer examination of the situation revealed that the instrument which had been used to measure mother's self-esteem had been tried and tested; this was not the case for the tool used to measure the children's self-esteem.

Answer: Some situations might involve more than one mistake. However, the most obvious are the following:

(1)   *SS*. Perhaps we should call this "reduced bias."
(2)   *HD*. The never-married people are most likely younger (perhaps teenagers) and would thus be healthier due to youth rather than marital status.
(3)   *IB*. Note that "agreed with her" implies that she shared her opinion with the men as she surveyed them.
(4)   *HD* and *PIB*. The differences in self-esteem scores may be attributed to noncomparable tools used for measuring adult and child self-esteem, rather than the children's self-esteem actually being significantly higher than the mothers.
Difficulty level: Easy to Medium

1.6.4: Consider the following list of statistical mistakes:

a. Failure to adjust to a per-item basis (PIB)
b. Forgetting to adjust for inflation (INF)
c. Induced bias (IB)
d. Self-selection (SS)
e. Hidden differences (HD)

In the situations below, indicate the type of mistake that was made. If no apparent mistake has been made, write "NM."

_____ (1) In order to compare the results of two sales approaches, a sales manager asked his sales people to use each of the approaches on half of their assigned customers. The approach to be used was determined by a coin flip prior to each sales call. The sales commission and the amount of training were the same.

_____ (2) It appears that the cost of supplying new tires for a fleet of taxis has gotten out of hand. Three years ago our company paid $130,000 for new tires, this year $275,000. Management has obviously been lax in monitoring this cost.

_____ (3) Records at an automobile insurance company show that pickup trucks owned by corporations are more likely to be involved in accidents than pickup trucks owned by private individuals. People therefore seem to be more careful when they drive their own vehicles.

_____ (4) It has been reported that in the first week after President Gerald Ford pardoned Richard Nixon the White House received 4,000 letters. Only 700 of the letters favored the pardon. The nation was obviously upset by Ford's action.

Answer: Some situations might involve more than one mistake. However, the most obvious are the following:

(1) *NM*. We might want to follow up this question and ask if there would be a difference if half the sales people were assigned to use each approach. Doing so would allow us to determine whether the differences in sales results were due to differences in approach or people.
(2) *INF* and *PIB*. The cost of the tires should be compared on a per-mile-driven basis, and the uncontrollable inflationary costs must also be taken into account.
(3) *PIB*. The accident rates per mile driven need to be compared. Commercial vehicles usually travel more miles and hence have greater exposure to accidents.
(4) *SS*. Only those citizens who were highly pleased or highly angered were likely to write. Thus, no generalization of "the nation" is possible.
Difficulty level: Easy to Medium

Empirical Frequency Distributions

2.1.1: Which of the following is a tabular or graphical format commonly used to organize and summarize data?

(a)   An array
(b)   An ogive
(c)   A stem-and-leaf diagram
(d)   A frequency distribution
(e)   All of these

Answer: (d)
Difficulty level: Easy

2.1.2: Which of the following graphical methods would be MOST useful for summarizing a large data set?

(a)   An array
(b)   A stem-and-leaf diagram
(c)   A frequency distribution
(d)   An array or a stem-and-leaf diagram
(e)   All of these

Answer: (e)
Difficulty level: Easy

2.1.3: Which of the following graphical methods are useful for identifying outliers in a data set?

(a)   An array only
(b)   A frequency distribution only
(c)   A box plot only
(d)   Either an array or a frequency distribution
(e)   Either an array or a box plot

Answer: (e)
Difficulty level: Easy to Medium

2.1.4: Which of the following DOES NOT represent a task required for constructing a frequency distribution?

(a)   Selecting the number of classes
(b)   Selecting the class boundaries
(c)   Determining the appropriate sample size
(d)   Choosing the class interval or width of the classes
(e)   Counting the number of values that fall in each class

Answer: (c)
Difficulty level: Easy

2.1.5: Assume that you have a data set that consists of the lengths of telephone conversations on 500 randomly selected long distance calls placed by employees of a government agency. What is the approximate number of classes needed to summarize these data using a frequency distribution?

(a)   5
(b)   10
(c)   15
(d)   20
(e)   8

Answer: (b)
Difficulty level: Easy

2.1.6: Forty students in a freshman English course were graded on their first assignment. The grades are:

80	78	56	93	84	76	70	50	45	60
73	73	85	95	60	55	76	81	65	72
50	75	61	89	93	81	75	79	96	50
76	81	91	50	73	75	71	43	80	90

(a)   How many students scored at least 80 on the assignment?
(b)   What percentage of the students scored at least 70 but less than 90 on the assignment?
(c)   Determine the approximate number of classes for the corresponding frequency distribution.
(d)   Determine the class interval or width.

Answer: a.   14
            b.   55%
            c.   5
            d.   11, starting with 42 and ending with 97
Difficulty level: Easy to Medium

2.1.7: Which of the following *best* describes the rules or procedures one uses to construct frequency distributions?

(a) Classes need not be mutually exclusive and exhaustive.
(b) There are well established rules that must be followed.
(c) The rules for selecting the number of classes and class widths are agreed to and followed by all statisticians.
(d) There are no universal rules for making choices required to construct frequency distributions.
(e) None of these.

Answer: (d)
Difficulty level: Easy to Medium

2.1.8: Which of the following statements about a frequency distribution is true?

(a) It is advantageous to vary the class width when constructing a frequency distribution.
(b) It is important to have as many classes as possible when constructing a frequency distribution.
(c) A frequency distribution can readily show the overall pattern of the data.
(d) Data values can fall into more than one class in a frequency distribution.
(e) None of these.

Answer: (c)
Difficulty level: Easy

2.1.9: Which of the following accurately describes the relationship between absolute frequency and relative frequency?

(a) Relative frequency represents absolute frequency divided by the number of observations in the data set.
(b) Absolute frequency represents relative frequency divided by the number of observations in the data set.
(c) Relative frequency represents the sum of observations that are less than a given value. Absolute frequency represents the number of observations in a class.
(d) Absolute frequency represents the sum of observations that are less than a given value. Relative frequency represents the number of observations in a class.
(e) Absolute frequency and relative frequency are identical ways of expressing grouped data.

Answer: (a)
Difficulty level: Easy

2.2.1: We can construct a cumulative distribution for:

(a)   A frequency distribution.
(b)   Relative frequency.
(c)   Percentages.
(d)   Absolute frequencies.
(e)   All of these.

Answer: (e)
Difficulty level: Easy

2.2.2: Which of the following describes the relationship between absolute frequency and cumulative frequency?

(a)   Cumulative frequency represents absolute frequency divided by the number of observations in the data set.
(b)   Absolute frequency represents cumulative frequency divided by the number of observations in the data set.
(c)   Cumulative frequency is the number of observations less than a given value, absolute frequency is the number of observations in a given class.
(d)   Cumulative frequency is the number of observations in a given class, absolute frequency is the percentage.
(e)   All of these.

Answer: (c)
Difficulty level: Easy

2.2.3: Use the data below to answer the following questions:

19	23	28	36	30	12	37	44	11	44
44	9	33	29	15	30	24	8	9	33
35	21	29	34	13					

n = 25

a. Construct an array.
b. Construct a cumulative frequency distribution.
c. What number of observations are less than 36?
d. What is the range for the above set of data?

Answer: a. 8,9,9,11,12,13,15,19,21,23,24,28,29,
            30,30,33,33,34,35,35,36,37,44,44
         b. A possible cumulative frequency distribution
            is given below:

Class Limits		Cumulative Frequency
6	- 15	7
16	- 25	11
26	- 35	21
36	- 45	25

         c. 21
         d. Range = (44 - 8) = 36

Difficulty level: Easy

2.3.1: Which of the following is true about histograms?

(a) They provide a pictorial representation of the
    information contained in a frequency distribution.
(b) The larger the class interval or width, the smaller the
    amount of information lost.
(c) They are constructed from cumulative frequency
    distributions.
(d) They greater the number of classes, the better the data
    are summarized by the histogram.
(e) They are constructed from ogives.

Answer: (a)
Difficulty level: Easy

2.3.2: Which of the following best describes the relation between a frequency distribution and a histogram?

(a) A histogram is constructed from a cumulative frequency distribution.
(b) A histogram is constructed from an absolute frequency distribution.
(c) A histogram is more appropriate for large data sets. A frequency distribution is more appropriate when using a small data set.
(d) A frequency distribution is more appropriate for large data sets. A histogram is more appropriate when using a small data set.
(e) Frequency distributions and histograms are methods of graphically presenting data that are related to one another in a way that changes with the application.

Answer: (b)
Difficulty level: Easy to Medium

2.4.1: Which of the following are *advantages* of using ogives to graphically present and summarize data?

(a) Can interpolate graphically.
(b) Represents a cumulative frequency distribution graphically.
(c) Can approximate percentiles when we have cumulative relative frequencies for each of the classes.
(d) Easy to determine approximate median value.
(e) All of these.

Answer: (e)
Difficulty level: Easy

2.5.1: Which of the following statements is *most* accurate about stem-and-leaf displays?

(a) Stem-and-leaf displays are better suited to working with large data sets than are histograms.
(b) Stem-and-leaf displays are generally used to summarize the information in vast amounts of business data.
(c) Stem-and-leaf displays display the actual data values whereas histograms and frequency distributions do not.
(d) Stem-and-leaf displays display the actual data values and are well suited to working with large data sets.
(e) Stem-and-leaf displays are better suited to working with small data sets than are histograms.

Answer: (c)
Difficulty level: Easy to Medium

2.5.2: Which of the following statements is (are) correct?

(a) Histograms and frequency distributions are useful for summarizing ratio data.
(b) Histograms and frequency distributions are generally used to summarize the information in vast amounts of business data.
(c) There is greater flexibility in selecting class intervals for histograms than in selecting stems for stem-and-leaf displays.
(d) Stem-and-leaf displays display the actual data values whereas histograms and frequency distributions do not.
(e) All of these.

Answer: (e)
Difficulty level: Easy to Medium

2.5.3: A(n) _____ presents the data values in a pictorial display to expose statistical outliers and patterns of data.

(a) ogive
(b) histogram
(c) stem-and-leaf diagram
(d) frequency distribution
(e) cumulative frequency distribution

Answer: (c)
Difficulty level: Easy

2.6.1: A(n) _____ is a list of the data with the numerical data ordered in either ascending or descending order.

(a) array
(b) ogive
(c) histogram
(d) frequency distribution
(e) cumulative frequency distribution

Answer: (a)
Difficulty level: Easy

2.6.2: Which of the following statements is (are) correct?

(a)   Histograms and frequency distributions are typically
       used to graphically summarize nominal and ordinal data.
(b)   Histograms and frequency distributions are typically
       used by managers who are working with large data sets.
(c)   Grouped bar charts are well suited for comparing two
       categories of continuous ratio data over time.
(d)   Stem-and-leaf displays are typically used by managers
       who are working with large data sets.
(e)   All of these.

Answer: (b)
Difficulty level: Easy to Medium

2.6.3: The *difference* between histograms and ogives is that:

(a)   ogives are only used for large data sets.
(b)   histograms are only used for large data sets.
(c)   there are no differences between a histogram and an
       ogive.
(d)   histograms are constructed from cumulative frequency
       distributions, ogives are constructed from absolute
       frequency distributions.
(e)   histograms are constructed from absolute frequency
       distributions, ogives are constructed from cumulative
       frequency distributions.

Answer: (e)
Difficulty level: Easy to Medium

2.6.4: A(n) is a tabular summary of a set of data that shows
the frequency or number of data items that fall in each of
several classes.

(a)   array
(b)   ogive
(c)   histogram
(d)   frequency distribution
(e)   cumulative frequency distribution

Answer: (e)
Difficulty level: Easy

2.6.5: Thirty middle-aged men were weighed as a part of an overall physical examination. The weights are given below:

173	145	155	150	230	192
166	160	167	205	140	177
169	170	148	193	153	198
228	195	158	165	214	180
160	172	173	203	145	167

a.   Construct a stem-and-leaf diagram for these weights. Comment on the shape of the distribution of weights for the men contained in the sample.
b.   Where do the data values seem to concentrate?
c.   How many of the male weights are less than 130 pounds?
d.   What percentage of the male weights exceed 150 pounds?

Answer: a.   The stem-and-leaf diagram below indicates a positively (or rightly) skewed distribution for the weights of the thirty men sampled.

```
14  |  0   5   5   8
15  |  0   3   5   8
16  |  0   0   5   6   7   7   9
17  |  0   2   3   3   7
18  |  0
19  |  2   3   5   8
20  |  3   5
21  |  4
22  |  8
23  |  0
```

b.   The data values seem to be concentrated around weights in between 160 - 170 pounds.
c.   8.
d.   9/30 = .30.

Difficulty level: Easy to Medium

2.6.6: Which of the following is *not* a graphical method used to summarize and present data?

(a)   A histogram.
(b)   A stem-and-leaf display.
(c)   A frequency distribution.
(d)   A cumulative frequency distribution.
(e)   All of these are graphical methods used to summarize and present data.

Answer: (e)
Difficulty level: Easy

2.6.7: Which of the following statements is (are) correct?

(a)   Stem-and-leaf displays have the advantage over histograms of displaying the actual data values.
(b)   Investors and managers often analyze and interpret financial statements using ratio analysis.
(c)   Histograms are often used to determine or assess the central tendency and/or dispersion of the data values.
(d)   Exploratory data analysis is useful for detecting and monitoring the existence of statistical outliers.
(e)   All of these.

Answer: (e)
Difficulty level: Easy

2.6.8: A(n) _____ is a graphic presentation of a frequency distribution whereas a(n) _____ is a graphic representation of a cumulative frequency distribution.

(a)   array, ogive
(b)   ogive, array
(c)   ogive, histogram
(d)   histogram, ogive
(e)   histogram, array

Answer: (d)
Difficulty level: Easy

Descriptive Measures

3.2.1: Which of the following are measures of central tendency or location?

(a)   Mean, median, and mode
(a)   Mean, median, and range
(b)   Mean, median, and variance
(c)   Mean, median and standard deviation
(e)   Range, variance, and standard deviation

Answer: (a)
Difficulty level: Easy

3.3.1: A personnel manager wants to estimate the average age of its sales people. Her sample of ages for ten salespeople is listed below.

    Age   46   49   32   30   27   49   62   53   37   39
(in years)

What are the mean and median ages of the salespeople who were sampled by the personnel manager?

(a)   42.4 years, 42.4 years
(a)   42.5 years, 42.5 years
(b)   42.4 years, 42.5 years
(c)   42.5 years, 42.4 years
(e)   The mean and median ages cannot be determined without
      further information being given

Answer: (b)
Difficulty level: Easy

3.3.2: A sample of delay times for Greyhound bus arrivals at a northeastern bus depot is listed below.

            6   6   4   8   10   10   12

The mean and median delay times are:

(a)   8.0, 8.0
(b)   9.3, 8.0
(c)   8.0, 6 and 10
(d)   9.3, 6 and 10
(e)   The mean and median delay times cannot be determined
      without further information being given

Answer: (a)
Difficulty level: Easy

3.4.1: A *weighted mean* is used:

(a) only when working with grouped data.
(b) when some of the observed values being averaged are of equal importance, but not all are.
(c) when all of the observed values being averaged are of equal importance.
(d) whenever we are interested in determining the relative variability of two data sets.
(e) none of these.

Answer: (b)
Difficulty level: Easy

3.4.2: A professor teaches two classes of statistics. On the first exam, his first class of 30 students had an average score of 90 and his second class of 20 students had an average score of 80. What is the *combined* mean for the two classes of students taking the professor's first exam?

(a) 80
(b) 84
(c) 86
(d) 90
(e) 170

Answer: (c)
Difficulty level: Easy

3.5.1: The procedure for computing the sample median:

(a) involves adding up all of the observations and then dividing by the sample size n.
(b) involves adding up all of the observations and then dividing by the term (n - 1).
(c) involves first constructing an array and then choosing the single middle value if there are an even number of data values.
(d) involves first constructing an array and then averaging the two middle values if there are an even number of data values.
(e) the median cannot be determined without knowledge of the exact value of the sample mean.

Answer: (d)
Difficulty level: Easy

3.5.2: The measure of central tendency, which by definition has the property of equally separating the upper and lower half of an arrayed data set, is the:

(a)   mean.
(b)   weighted mean.
(c)   median.
(d)   mode.
(e)   range.

Answer: (c)
Difficulty level: Easy

3.6.1: A sample of delay times for 11 people waiting for airline arrivals at O'Hare Airport in Chicago are listed below. What are the median and modal waiting times for airline arrivals?

Waiting time   20 16 81 19 20 66 16 13 52 77 26
(in minutes)

(a)   38.6 minutes, 20 minutes
(b)   20 minutes, 38.6 minutes
(c)   20 minutes, 16 and 20 minutes
(d)   16 and 20 minutes, 20 minutes
(e)   16 and 20 minutes, 38.6 minutes

Answer: (c)
Difficulty level: Easy

3.6.2: The data values below represent the number of people supervised by 6 department heads in a department store. What are the median and modal number of people supervised?

          5   5   12   11   8   7

(a)   8.0, 8.0
(b)   8.0, 7.5
(c)   8.0, 5.0
(d)   7.5, 8.0
(e)   7.5, 5.0

Answer: (e)
Difficulty level: Easy

3.6.3: The median and mode:

(a) are both measures of dispersion or scatter.
(b) are both measures of central tendency or location.
(c) are resistant to statistical outliers in the data.
(d) always better measure central tendency than the mean.
(e) always use every value of a data set.

Answer: (c)
Difficulty level: Easy

3.6.4: Which of the following is an *advantage* of using
the mean instead of the median or mode as a measure of
central tendency or location?

(a) The mean is efficient; it uses every data value.
(b) The mean is resistant to outlying values.
(c) The value of the mean is often cumbersome to calculate
     and difficult to interpret.
(d) The mean has the property of equally dividing the data
     set into two equal halves.
(e) None of these are advantages of using the mean instead
     of either the median or the mode.

Answer: (a)
Difficulty level: Easy to Medium

3.6.5: Which of the following is an *advantage* of using the
mode instead of the mean as a measure of location?

(a) There may be more than one mode.
(b) The mode is more widely used than the mean.
(c) The exact value of the mode can be determined from a
     frequency distribution.
(d) The mode is resistant to outlying values.
(e) The mode has the property of equally dividing the data
     set into two equal halves.

Answer: (d)
Difficulty level: Easy

3.6.6: Which of the following is an *advantage* of using the median instead of the mean as a measure of location?

(a)   The median uses every data value.
(b)   The median has the property of equally dividing the data set into two equal halves.
(c)   The median is more widely accepted and used by statisticians than either the mean or the mode.
(d)   The median has desirable mathematical properties that neither the mean nor the mode have.
(e)   None of these are advantages of using the median instead of either the mean or the mode.

Answer: (b)
Difficulty level: Easy

3.7.1: The standard deviation gives us:

(a)   a typical or representative value for a data set.
(b)   a standard measure for comparison.
(c)   a measure of variation or scatter.
(d)   a measure of variation with the same units as the data.
(e)   all of these.

Answer: (c)
Difficulty level: Easy

3.8.1: Which of the following are reasons why the range is a *poor* measure of variation?

(a)   As the number of observations is increased, the range generally tends to become larger.
(b)   Its calculation involves only two of the observed values regardless of the number of observations.
(c)   The range is the least stable measure of variation for all but the smallest sample sizes.
(d)   The range lacks desirable mathematical properties.
(e)   All of these are reasons why the range is a poor measure of variation.

Answer: (e)
Difficulty level: Easy

3.8.2: Which of the following are reasons why the sample variance is *better* than the range as measuring the amount of variability or variation in a data set?

(a)   The variance is easier to compute than the range.
(b)   The variance has desirable mathematical properties that the range does not.
(c)   The variance is always expressed in the same units as the (values in the) data set.
(d)   The variance is easy to interpret using the Empirical Rules for standard deviations.
(e)   All of these are reasons why the variance provides a better measure of variation than the range.

Answer: (b)
Difficulty level: Easy

3.9.1: The algebraic sum of the deviations of a set of numbers from their mean:

(a)   may or may not equal zero
(b)   cannot possibly equal zero
(c)   is a good measure of variation that is of great interest to statisticians
(d)   is a poor measure of variation since it equals zero
(e)   all of these

Answer: (d)
Difficulty level: Easy

3.10.1: A professor teaches two sections of a statistics class, each of which have 20 students. On the professor's first exam, the students in the first class all received a score of 70. The mean score in the professor's second class was 70 with a variance of 10. What is the variance of the *combined* class, treating each class as a population?

(a)   2.5
(b)   5
(c)   10
(d)   20
(e)   0

Answer: (b)
Difficulty level: Challenging

29

3.10.2: How are the variance and the standard deviation related mathematically?

(a) The variance is always twice the standard deviation.
(b) The variance squared equals the standard deviation.
(c) The standard deviation equals the positive square root of the variance.
(d) The standard deviation equals the variance.
(e) None of these.

Answer: (b)
Difficulty level: Easy

3.10.3: If the population variance of a set of data is 144, the standard deviation is:

(a) 72
(b) 12
(c) 20,736
(d) 36
(e) unable to be determined without additional information

Answer: (b)
Difficulty level: Easy

3.10.4: Which of the following *cannot* be a numerical value for the variance of a set of data?

(a) 1050
(b) 42
(c) 0
(d) -144
(e) All of these are acceptable values for a variance

Answer: (d)
Difficulty level: Easy

3.11.1: For the list of variables below, which *formulas* would produce the same numerical value for both the population and the sample?

(a) Mean
(b) Variance
(c) Standard deviation
(d) Coefficient of variation
(e) None of these

Answer: (a)
Difficulty level: Easy to Medium

3.11.2: A sample of 10 waiting times for airline arrivals at O'Hare Airport in Chicago are listed below.
Waiting time   9   6   6   4   8   10   10   12   15   40
(in minutes)

a.   Compute the sample mean.
b.   Determine the median and mode(s).
c.   Compute the sample variance.
d.   Compute the sample standard deviation.

Answer: a.   Mean, $\overline{X}$ = (1/10)[9 + ... + 40] = 12.0 minutes
                          = (1/10)[120] = 12.0 minutes
           b.   Median, Md = (1/2)[9 + 10] = 9.5 minutes
                 Mode(s), Mo = 6 and 10 minutes
           c.   Variance, s^2 = (1/9)[(9-12)^2 + ... + (40-12)^2]
                          = (1/9)[962] = 106.9 (minutes)^2
           d.   Standard deviation, s = square root of s^2
                          = 10.3 minutes
Difficulty level: Easy to Medium

3.11.3: Explain in your own words *why* it is necessary to divide by (n - 1) when calculating the sample standard deviation, when the calculation for the population standard deviation uses n as the divisor.

Answer: We divide by (n - 1) because we lose one degree of
           freedom when calculating the sample variance,
           since the range of the sample is smaller than the
           range of the population.
Difficulty level: Easy to Medium

3.12.1: A random sample of average annual salaries for 50 college graduates majoring in economics is listed below.
     Mean = $36,000
     Standard deviation = $6,000

Given that this distribution is symmetric and bell-shaped, what is the approximate percentage of college graduates majoring in economics with annual salaries in between $24,000 and $48,000?

(a)   It is approximately 95%.
(b)   It must be at least 75%.
(c)   It is approximately 5%.
(d)   It is at most 25%.
(e)   It is approximately zero.

Answer: (a)
Difficulty level: Easy to Medium

3.12.2: A mechanical bolt cutter cuts bolts with an average diameter of 0.51 inches and a standard deviation of 0.02 inches. Bolt specifications require that bolts have a diameter equal to 0.50 ± 0.05 inches so that bolts that do not satisfy this requirement are considered defective. If the distribution of bolt diameters is approximately normal, what approximate fraction of total production would result in *defective* bolts?

(a) 2.41%
(b) 4.82%
(c) 97.59%
(d) 95.18%
(e) not enough information is given to solve the problem

Answer: (a)
Difficulty level: Medium to Challenging

3.12.3: Suppose you knew that the population of I.Q.'s for people in the eastern United States had a mean of 115 and a standard deviation of 15. Use the General Empirical Rule or the Rule for Bell-Shaped Data where appropriate to determine the following:

a. What maximum percentage of people in the eastern United States could possibly have I.Q.'s below 85?
b. What maximum percentage of people in the eastern United States could possibly have I.Q.'s that are either below 85 or above 145?
c. What minimum percentage of people in the eastern United States must have I.Q.'s in between 85 and 145?
d. What minimum percentage of people in the eastern United States must have I.Q.'s in between 70 and 160?

Answer: a. $1/(2)^2 = 1/4 = .25$.
          b. $1/(2)^2 = 1/4 = .25$.
          c. $[1 - 1/(2)^2] = [1 - 1/4] = 3/4 = .75$.
          d. $[1 - 1/(3)^2] = [1 - 1/9] = 8/9 = .889$.
Difficulty level: Easy to Medium

3.12.4: A manufacturer of computer hard disk drives knows that its computer hard disk drives have a mean time to failure of 24 months with a standard deviation of 4 months. Use the General Empirical Rule or the Rule for Bell-Shaped Data where appropriate to determine the following:

a.   What is the minimum proportion of the manufacturers' computer hard disk drives that will not fail earlier than 12 months?
b.   What is the minimum proportion of the manufacturers' computer hard disk drives that will fail in between 16 and 32 months?
c.   If the manufacturer offers a one year (12 month) warranty on its' computer hard disk drives, what is the maximum percentage of these hard drives that could possibly fail while under warranty?
d.   What warranty length (to the nearest month) should the manufacturer choose so that a maximum of only 4% (1/25) of the computer hard disk drives could possibly fail while still under warranty?
e.   Explain how your answers would change when you assume that the distribution of computer hard disk drives were normally distributed.

Answer: a. $[1 - (1/3)^2] = [1 - 1/9] = 8/9 = .889$.
           b. $[1 - (1/2)^2] = [1 - 1/4] = 3/4 = .75$.
           c. $(1/3)^2 = 1/9 = .111$.
           d.   First solve for k, then solve for the number of months that you will warranty the disk drives. We first set $(1/k)^2 = 1/25$. Solving for k yields: $k = 5$. Solving for X yields: $X = 4$. Our warranty should be for 4 months.
           e.   Since the normal distribution is bell-shaped, we may use the rule for bell-shaped data. For parts a-c, the exact answers are .9987, .9544, and .0013, although these can be roughly approximated by 100%, 95%, and 0%, using the rules for bell-shaped data. Our answer for part d is more difficult, as it requires the use of the standard normal or z table. We are looking for that z value, call it b, with the property that $P(z < b) \le .04$. We therefore look up an area in the z table that is just above .4600. The z value that does this is -1.76. Our answer to part d can be found using the formula: $x = \mu + z\sigma$. Solving for x yields: $x = 24 + (-1.76)(4) = 16.96$, so we should only warrantee the computer hard disk drives for 16 months.
       Difficulty level: Easy to Challenging

3.12.5: Suppose you knew that the population of I.Q.'s for people in the eastern United States was bell-shaped and symmetric with a mean of 115 and a standard deviation of 15. Use the General Empirical Rule or the Rule for Bell-Shaped Data where appropriate to determine the following:

a. What percentage of people in the eastern United States have I.Q.'s below 85?
b. What percentage of people in the eastern United States have I.Q.'s that are either below 85 or above 145?
c. What percentage of people in the eastern United States have I.Q.'s in between 85 and 145?

Answer: a.   Approximately .025.
          b.   Approximately .050.
          c.   Approximately .950.
Difficulty level: Easy to Medium

3.12.6: If we have a data set where the histogram reveals a distribution that is skewed right, which of the following statements *must* be true?

(a) At least 89% of the values must lie within $\mu \pm 3\sigma$.
(b) Approximately 95% of the values will lie within $\mu \pm 2\sigma$.
(c) Approximately 68% of the values will lie within $\mu \pm 1\sigma$.
(d) At most 75% of the values could possibly lie outside of $\mu \pm 2\sigma$.
(e) All of these.

Answer: (a)
Difficulty level: Medium

3.13.1: The 1990 Census Bureau reported that for a small town of 1,000 families located in the eastern United States, the mean and median annual incomes were $30,000 and $28,000. The *total* annual income for the whole town:

(a) is $280,000.
(b) is $300,000.
(c) is $2,800.
(d) is $3,000.
(e) cannot be determined without additional information.

Answer: (b)
Difficulty level: Easy

3.13.2: If a distribution is symmetric and unimodal, then:

(a)   the mean, median, and mode are all equal.
(b)   the distribution is said to be skewed.
(c)   the mean is the largest of the 3 values.
(d)   the mean is the smallest of the 3 values.
(e)   the mean lies in between the median and the mode.

Answer: (a)
Difficulty level: Easy

3.13.3: The 1990 Census Bureau reported that for a small
town of 1,000 families, located in an eastern state, the
mean and median annual incomes were $30,000 and $28,000.
From this information alone, we know that:

(a)   the distribution of annual family incomes in this town
      is symmetric and a possible value for the modal income
      is $30,000.
(b)   the distribution of annual family incomes in this town
      is skewed right and a possible value for the modal
      income is $26,000.
(c)   the distribution of annual family incomes in this town
      is skewed left and a possible value for the modal
      income is $26,000.
(d)   the distribution of annual family incomes in this town
      is skewed right and a possible value for the modal
      income is $32,000.
(e)   the distribution of annual family incomes in this town
      is skewed left and a possible value for the modal
      income is $32,000.

Answer: (b)
Difficulty level: Easy to Medium

3.13.4: Assume that the distribution of income among people
living in a suburban area is skewed right. A real estate
manager who wishes to emphasize the *affluence* of the
suburban area would likely quote the _____ family income
to prospective home buyers.

(a)   mean
(b)   modal
(c)   median
(d)   lowest decile
(e)   lowest quartile

Answer: (a)
Difficulty level: Easy

3.14.1: The *coefficient of variation* expresses:

(a)   the standard deviation as a percentage of the mean.
(b)   the mean as a percentage of the standard deviation.
(c)   the median as a percentage of the mean.
(d)   the range as a percentage of the variance.
(e)   the variance as a percentage of the mean.

Answer: (a)
Difficulty level: Easy

3.14.2: A measure of relative variation expressed in percentage terms is:

(a)   the mean
(b)   the coefficient of variation
(c)   the sum of squares
(d)   the standard deviation
(e)   none of these

Answer: (b)
Difficulty level: Easy

3.15.1: *Exploratory data analysis* is most useful for:

(a)   graphically presenting continuous ratio data.
(b)   graphically presenting discrete nominal data.
(c)   identifying statistical outliers and exposing patterns of data.
(d)   better understanding the average value and amount of variability in a data set.
(e)   better understanding the relationship between two or more variables of interest.

Answer: (c)
Difficulty level: Easy

3.15.2: Which of the following are terms (or tools) that are associated with exploratory data analysis?

(a)   histograms and box plots
(b)   histograms and stem-and-leaf diagrams
(c)   histograms and frequency distributions
(d)   box plots and pie charts
(e)   box plots and stem-and-leaf diagrams

Answer: (e)
Difficulty level: Easy

3.15.3: Construct a stem-and-leaf display for the data below that represents a sample of 40 weights of olympic swimmers:

```
150   160   168   172   178   181   184   188   192   196
154   158   162   167   171   177   183   186   188   192
154   166   169   172   174   182   186   188   191   195
158   168   169   170   177   179   185   186   189   203
```

Answer: There are many possibilities.   Below is one:

```
15 | 0   4   4   8   8
16 | 0   2   6   7   8   8   9   9
17 | 0   1   2   2   4   7   7   8   9
18 | 1   2   3   4   5   6   6   6   8   8   8   9
19 | 1   2   2   5   6
20 | 3
```

Difficulty level: Easy

3.16.1: Which of the following is a frequency distribution?

(a)   the number of faculty members at New York University classified by subject area.
(b)   the number of car owners who commute using the New York subway classified by annual income.
(c)   the number of college students who commute to a local community college.
(d)   the number of lawyers in Washington, D.C. classified by the law school they received their degrees.
(e)   the number of people waiting to depart on flights going out of O'Hare Airport classified by name.

Answer: (b)
Difficulty level: Easy to Medium

3.16.2: Which of the following is a frequency distribution?

(a)   the number of students attending Boston University classified by numerical grade point average.
(b)   the number of automobiles in Massachusetts classified by color.
(c)   the number of registered automobiles classified by state.
(d)   the number of lobbyists in Washington, D.C. classified by gender.
(e)   the number of teachers in public schools classified by level being taught by the teacher.

Answer: (a)
Difficulty level: Easy to Medium

3.16.3: During a recent labor negotiation between a teachers union and a school board, representative wages were discussed. Frequently the teachers union cites one measure and the school board cites another, numerically different, measure of the representative wage. Given your knowledge of central tendency, it would be most logical to conclude that:

(a)   the teachers union will use the mean wage to make its point while the school board will cite the median wage.
(b)   the teachers union will use the median wage to make its point while the school board will cite the mean wage.
(c)   the teachers union will use the modal wage to make its point while the school board will cite the mean wage.
(d)   the teachers union will use the mean wage to make its point while the school board will cite the median wage.
(e)   the teachers union will use the modal wage to make its point while the school board will cite the median wage.

Answer: (c)
Difficulty level: Easy to Medium

3.16.4: Which of the following accurately describes the differences between the General Empirical Rule and the Rule for Bell-Shaped Data?

(a)   The General Empirical Rule is more widely applicable and more precise than the Rule for Bell-Shaped Data.
(b)   The General Empirical Rule is less widely applicable but more precise than the Rule for Bell-Shaped Data.
(c)   The General Empirical Rule is more widely applicable but less precise than the Rule for Bell-Shaped Data.
(d)   The General Empirical Rule is less widely applicable and less precise than the Rule for Bell-Shaped Data.
(e)   None of these.

Answer: (c)
Difficulty level: Easy to Medium

3.16.5: A sample of money market mutual fund annual yields is listed below.

6   6   4   8   10   10   12

a.   Compute or find the mean and median annual yields.
b.   Compute the variance and standard deviation.

Answer: a.   Mean = 8, Median = 8
          b.   Variance = 8, Standard deviation = 2.8
Difficulty level: Easy

3.16.6: Fifteen applicants to an M.B.A. program took the
GMAT, a test required for admission to the program. The
scores made are given below:

320	400	290	510	550
230	490	260	550	480
280	360	280	420	370

a.   Calculate the mean of this population.
b.   Calculate the median.
c.   Calculate the mode.
d.   Calculate the range.
e.   Calculate the standard deviation of this population.
f.   Calculate the coefficient of variation.

Answer: a.   386
           b.   370
           c.   280, 550
           d.   320
           e.   105.63
           f.   27.4%
Difficulty level: Easy to Medium

3.17.1: The frequency distribution below shows the annual
salaries of a sample of 50 financial analysts working in a
region of the eastern United States.

Annual Salary (in $1,000's)	Midpoint (in $1,000's)	Frequency (# analysts)
30 less than 50	40	10
50 less than 70	60	20
70 less than 90	80	10
90 less than 110	100	5
110 less than 130	120	5

a.   The approximate mean and median.
b.   The variance and standard deviation.
c.   The number of analysts having annual salaries of at
     least $50,000 but less than $90,000.

Answer: a.   Mean = $70,000, Median = $65,000
           b.   Variance = 5.918×10^8 (squared dollars),
                 Standard deviation = $24,327.69
           c.   Frequency = (20 + 10) = 30 analysts
Difficulty level: Easy to Medium

3.17.2: The frequency distribution below shows a sample of annual percentage returns on investment portfolios chosen by 50 investment managers from the eastern United States.

Annual Return (in %)	Midpoint (in %)	Frequency (# portfolios)
0 less than 6	3	5
6 less than 12	9	10
12 less than 18	15	20
18 less than 24	21	10
24 less than 30	27	5

a.   The approximate mean and median.
b.   The variance and standard deviation.
c.   Comment on the shape of the frequency distribution.

Answer: a.   Mean = 15%, Median = 15%.
           b.   Variance = 44.08 (squared percent), Standard
                 deviation = 6.64%.
           c.   It is approximately unimodal and symmetric.
Difficulty level: Easy to Medium

3.17.3: The relative frequency distribution below shows the annual salaries for a sample of 32 accountants working in a region of the eastern United States.

Annual Salary (in $1,000's)	Midpoint (in $1,000's)	Relative Frequency (% accountants)
50 less than 70	60	.500
70 less than 90	80	.250
90 less than 110	100	.125
110 less than 130	120	.125

a.   The approximate mean and the modal class.
b.   The variance and standard deviation.
c.   Comment on the shape of the frequency distribution.

Answer: a.   Mean = $77,500, Modal class $50,000 < $70,000
           b.   Variance = 4.58×10^8 (squared dollars),
                 Standard deviation = $21,402.44
           c.   It is skewed to the right (mean > median)
Difficulty level: Easy to Medium

3.17.4: The frequency distribution below shows the IQ's for a sample of 32 high school graduates from public schools in a region of the southern United States.

Student's IQ (IQ points)	Midpoint (IQ points)	Frequency (# graduates)
80 less than 100	90	6
100 less than 120	110	12
120 less than 140	130	8
140 less than 160	150	6

a. The approximate mean and median.
b. The variance and standard deviation.
c. Construct an interval for IQ's that is two standard deviations around the approximate sample mean IQ.

Answer: a.   Mean = 118.75, Median = 116.67.
           b.   Variance = 411.29, Standard deviation = 20.28.
           c.   118.75 ± (2)(20.28) or [78.19 to 159.31].
Difficulty level: Easy to Medium

3.17.5: The frequency distribution below shows the annual salaries for a sample of 50 lawyers working in a region of the midwestern United States.

Annual Salary (in $1,000's)	Midpoint (in $1,000's)	Frequency (# of lawyers)
20 less than 30	25	5
30 less than 40	35	25
40 less than 50	45	10
50 less than 100	75	5
100 less than 200	150	5

a. The approximate mean and the modal class.
b. The standard deviation and coefficient of variation.

Answer: a.   Mean = $51,500, Modal class is $30,000 less than $40,000.
           b.   Standard deviation = $35,574.73, Coefficient of variation = 0.691.
Difficulty level: Easy to Medium

3.17.6: The frequency distribution below shows the annual revenues for 1991, based on a sample of 32 pharmaceutical companies located in midwestern and eastern regions of the United States.

Annual Revenues (in millions)	Midpoint (in millions)	Frequency (# companies)
0 less than 5	2.5	12
5 less than 10	7.5	12
10 less than 20	15	4
20 less than 50	35	4

a.   The approximate mean and median.
b.   The variance and standard deviation.

Answer: a.   Mean = 10, Median = 6.67.
          b.   Variance = 1.08×10^{14} (squared dollars),
               Standard deviation = $10,395,408.41.
Difficulty level: Easy to Medium

3.17.7: The cumulative frequency distribution below shows the quantitative Scholastic Aptitude Test (SAT) scores for a sample of 1,500 students from colleges located in the eastern United States.

Quantitative SAT Score (in points, 0-800 scale)	Midpoint (in points)	Cumulative Frequency (# students)
0 less than 400	200	50
400 less than 500	450	500
500 less than 600	550	800
600 less than 700	650	1250
700 less than 800	750	1485
800 less than 900	850	1500

a.   The approximate mean and median.
b.   The standard deviation and coefficient of variation.

Answer: a.   Mean = 572.67 points, Median = 583.33 points.
          b.   Standard deviation = 129.41 points,
               Coefficient of variation = 0.226.
Difficulty level: Medium

3.17.8: The cumulative frequency distribution below shows the proportion of defective manufacturing parts produced by 50 lots at an industrial company in a region of the eastern United States.

Defective Parts (in %)	Midpoint (in %)	Cumulative Frequency (# of lots)
0 less than 2	1	15
2 less than 6	4	35
6 less than 10	8	45
10 less than 20	15	49
20 less than 100	60	50

a.   The approximate mean and median.
b.   The variance and standard deviation.

Answer: a.   Mean = 5.9%, Median = 4.0%.
          b.   Variance = 76.21 (squared percent),
               Standard deviation = 8.73%.
Difficulty level: Medium

3.17.9: The accompanying table gives the distribution of ages for the population of married women in the labor force of the eastern United States. Determine the following:

Age (in years)	# Married Women (in thousands)	Age (in years)	# Married Women (in thousands)
15-19	2212	40-44	3005
20-24	3503	45-54	5096
25-29	1313	55-64	2904
30-34	939	65-80	886
35-39	3490		

a.   What is the approximate median age of married women in the labor force who live in the eastern United States?
b.   What is the modal class for ages of married women in the labor force who live in the eastern United States?
c.   What approximate percentage of *single* women in the labor force live in the eastern United States?

Answer: a.   Approximate median age = 40.36 years.
          b.   Modal class is: 45-54 years old.
          c.   We cannot make any inferences regarding this approximate percentage because no such information on single women is provided.
Difficulty level: Easy to Medium

Probability

4.1.1: When using the _____ method for assigning relative frequency probabilities, the probability of an event is equal to the number of times that the event occurs in a large number of repetitions of the experiment divided by the total number of repetitions in the experiment.

(a) relative frequency
(b) classical
(c) independent
(d) mutually exclusive
(e) subjective

Answer: (a)
Difficulty level: Easy

4.1.2: A(n) _____ probability is an individual's degree of belief in the occurrence of an event.

(a) a priori
(b) classical
(c) independent
(d) mutually exclusive
(e) subjective

Answer: (e)
Difficulty level: Easy

4.1.3: A sample space is generally defined to be:

(a) the set of all equally likely outcomes for an experiment
(b) the set of all possible outcomes for an experiment
(c) the probability that an event will occur
(d) equal to one only for exhaustive events
(e) none of these

Answer: (b)
Difficulty level: Easy

4.2.1: Which of the following statements are FALSE?

(a)   if $\overline{A}$ is the complement of A, then $P(\overline{A}) = 1 - P(A)$.
(b)   if event A is independent of event B, then
      $P(A \text{ or } B) = P(A) + P(B)$ because of the fact that
      $P(A|B) = P(A)$ and $P(B|A) = P(B)$.
(c)   if event A is mutually exclusive of event B, then
      $P(A \text{ or } B) = P(A) + P(B)$ because of the fact that
      $P(A \text{ and } B) = 0$.
(d)   the sample space for a conditional probability is
      restricted to including only those outcomes where the
      given event has already occurred.
(e)   if A and B are exhaustive, then $P(A \text{ or } B)$ must equal 1.

Answer: (b)
Difficulty level: Easy to Medium

4.2.2: Which of the following is *not* a property to
which probabilities must conform?

(a)   If $\overline{A}$ is the complement of A, then $P(\overline{A}) = 1 - P(A)$.
(b)   If A is an event, then $0 \le P(A) \le 1$.
(c)   If event A is mutually exclusive of event B, then
      $P(A \text{ or } B) = P(A) + P(B)$.
(d)   The sample space contains all possible outcomes of the
      experiment. Thus $P(S) = 1$.
(e)   All of these are properties to which probabilities must
      conform.

Answer: (e)
Difficulty level: Easy to Medium

4.2.3: The general rule for addition of probabilities is
best summarized by the equation:

(a) $P(\overline{A}) = 1 - P(A)$.
(b) $P(A \text{ or } B) = P(A) + P(B) - P(A \text{ and } B)$.
(c) $P(A \text{ and } B) = P(A)P(B)$, if A and B are independent
      events.
(d) $P(A \text{ or } B) = P(A) + P(B)$, if A and B are mutually
      exclusive events.
(e) $P(A|B) = \dfrac{P(A)P(B|A)}{P(B)}$ for Bayes' theorem.

Answer: (b)
Difficulty level: Easy

4.3.1: If events A and B are independent events, then:

(a)   P(A and B) = 0
(b)   P(A or B) = 1
(c)   P(A and B) = P(A)P(B) = 0
(d)   P(A|B) = P(A) and P(B|A) = P(B)
(e)   Events A and B must also be mutually exclusive

Answer: (b)
Difficulty level: Easy

4.3.2: Event A is twice as likely to occur than event B. If
the two events are independent and P(A and B) = .32, then it
must be true that:

(a)   P(A) = 0.40 and P(B) = 0.80
(b)   P(B) = 0.40 and P(A) = 0.80
(c)   A and B are exhaustive events
(d)   A and B are mutually exclusive events
(e)   None of these must be true

Answer: (b)
Difficulty level: Easy to Medium

4.3.3: Event A is twice as likely to occur than event B and
events A and B are mutually exclusive. It *must* be true
that:

(a)   P(A and B) = P(A)P(B)
(b)   P(B) = 1/3 and P(A) = 2/3
(c)   A and B are exhaustive events
(d)   A and B are independent events
(e)   None of these must be true

Answer: (e)
Difficulty level: Easy to Medium

4.3.4: If P(A) = 0.5 and P(B) = 0.7, then which of the
following is a possible value for P(A and B) <u>given that</u> A
and B are not exhaustive events?

(a)   0
(b)   .30
(c)   .20
(d)   .55
(e)   .60

Answer: (b)
Difficulty level: Medium

4.3.5: If events A and B are mutually exclusive and exhaustive, then it *must* be true that:

(a)   P(A and B) = 0
(b)   P(A or B) = 0
(c)   P(A and B) = P(A)P(B)
(d)   P(A¦B) = P(A) and P(B¦A) = P(B)
(e)   P(A and B) = P(A) + P(B) = 1

Answer: (a)
Difficulty level: Easy

4.3.6: If two events A and B are given to be *statistically independent*, then we know that:

(a)   P(A or B) = 1
(b)   P(A and B) = 0
(c)   P(A and B) = 1
(d)   P(A and B) = P(A)P(B)
(e)   they must also be mutually exclusive

Answer: (d)
Difficulty level: Easy

4.5.1: For the night before the general election NBC has a fixed number of program breaks within which commercials can be purchased. Clinton's campaign has purchased time within 40% of these slots. Given that President Bush's campaign has purchased time in a particular slot, the probability that the slot will also have a Clinton commercial is 0.50. The probability that Clinton will have a commercial, given that Bush's campaign has *not* purchased a time in a particular slot is 0.30. The proportion of all NBC slots that have neither a Bush nor a Clinton commercial is:

(a)   0.64
(b)   0.48
(c)   0.35
(d)   0.24
(e)   0.12

Answer: (c)
Difficulty level: Easy to Medium

4.5.2: Assume that in the past a drug test for a highly addictive drug has correctly identified drug users 98% of the time and incorrectly identified non-users only 1% of the time. A confidential survey revealed that 10% of the workers at a unionized production facilities used this highly addictive drug.

a. What is the conditional probability that a person is a drug user, given that the drug test result is positive?
b. What is the conditional probability that a drug test mistakenly identifies the person as a non-user?
c. What percentage of the drug tests come back positive?

Answer: a.   .098/.107 = .9159 or 91.59%
           b.   .002/.893 = .0022 or .22%
           c.   .107 or 10.7%
Difficulty level: Easy to Medium

4.5.3: In a particular class with 50% males and 50% females, 4% of males received a final grade of an A and 16% of females received a final grade of an A. If a randomly selected person got an A, what is the conditional probability that the person is female?

(a)   .04
(b)   .25
(c)   .50
(d)   .75
(e)   .80

Answer: (e)
Difficulty level: Easy

4.5.4: It is known that 70% of all student applicants having a cumulative grade point average of at least 3.00 are admitted to a desired business program. Only 20% of all student applicants with cumulative grade point averages less than 3.00 are admitted to the program. If 60% of the student applicants for the program have cumulative grade point averages of at least 3.00, what percentage of the students are admitted to the business program?

(a)   42%
(b)   50%
(c)   74%
(d)   60%
(e)   Cannot be determined without further information

Answer: (b)
Difficulty level: Easy to Medium

4.6.1: A table summarizing the concentrations or majors for a survey of 250 students attending a business school in a region of the eastern United States is listed below.

	Accounting	Economics	Finance	Marketing	Totals
Female	20	10	50	30	110
Male	35	20	40	45	140
Totals	55	30	90	75	250

a. What is the probability that a student is a female accounting major?
b. What is the conditional probability that a student is a male, given that the student is a marketing major?
c. What is the probability that a student is either a female or a finance major, *but not both*?
d. What is the probability that a student is neither a male nor an economics major?
e. What is the conditional probability that a student is a finance major, given that they are male?

Answer: a.   20/250 = .080
          b.   45/75 = .600
          c.   100/250 = .400
          d.   100/250 = .400
          e.   40/140 = .2857
Difficulty level: Easy to Medium

4.6.2: Forty percent of the faculty of Union University hold a doctoral degree. Forty percent of Union's faculty who hold a doctorate are female.

a. Find the joint probability that a randomly selected faculty member is both a female and the holder of a doctoral degree.
b. What is the conditional probability that a faculty member holds a doctorate, given that the faculty member who is selected is male?
c. What is the probability that a faculty member who is selected at random does not hold a doctorate?

Answer: a.   .16
          b.   .60
          c.   .60
Difficulty level: Easy to Medium

4.6.3: A batch of transistors contains 10% that are
defective. Three transistors are drawn at random one at a
time, each being replaced prior to the draw. What is the
probability of obtaining *at least* one defective transistor?

(a)   .729
(b)   .001
(c)   .271
(d)   .750
(e)   none of the above

Answer: (c)
Difficulty level: Easy to Medium

4.6.4: Company records show that last month 70% of the
employees had used up all their vacation time. Half of
those people reportedly used some of their sick leave time
during the month. However, only 10% of the people who had
not used up their vacation time used sick leave last month.

a.   What proportion of the employees still had vacation
      time due them during last month?
b.   Find the conditional probability that a person did not
      use sick leave, given that they had vacation coming.
c.   Find the probability that a randomly selected person
      had vacation coming and took no sick leave.
d.   Find the probability that a person selected at random
      took sick leave last month.

Answer: a. $1.00 - .70 = .30$
           b. $1.00 - .10 = .90$
           c. $(.30)(.90) = .27$
           d. $(.70)(.50) + (.30)(.10) = .35 + .03 = .38$
Difficulty level: Easy to Medium

4.6.5: Three workers manufacture a part to be used in the
brake mechanism of new automobiles. The probability of each
of the workers making an error are .02, .01, and .06.
Worker A makes 45% of the parts, worker B makes 35% of the
parts, while worker C makes the remaining 20% of the parts.
The probability the next manufactured part is *defective* is:

(a)   .009
(b)   .991
(c)   .0035
(d)   .0245
(e)   .9755

Answer: (d)
Difficulty level: Medium

4.6.6: A utility burns coal from two sources--National Coal and General Coal. It gets about 65% of its coal shipments from National and the rest from General. Forty percent of National's coal is the low-sulfur variety, while only 20% of General's coal is the low-sulfur variety.

a. What is the probability that a randomly selected shipment is both low-sulfur and comes from General Coal?
b. A shipment of coal was recently discovered to have come from coal cars that were only partially full. Given that the coal was not of the low-sulfur type, what is the conditional probability that this shipment came from National Coal? from General Coal?

Answer: a.   P(Low sulfur and General) = (.35)(.20) = .07
          b.   P(National|Not low sulfur) = .39/[.39+.28]
                                          = .582
               P(General| Not low sulfur) = .28/[.39 + .28]
                                          = .418
Difficulty level: Easy to Medium

4.6.7: A major university's student population consists of 54% male students and 46% female students. In addition, 15% of all students are business students. A student is chosen at random. Assuming independence, find the probability that:

a.   the student is a male business student.
b.   the student is neither a male nor a business student.
c.   the student is either a female or a business student.
d.   the student is either a male or a business student but not both.

Answer: a.   .081
          b.   .391
          c.   .541
          d.   .528
Difficulty level: Easy to Medium

4.6.8: If P(A) = .30, P(B) = .70, and P(A and B) = .15, then it *must* be true that:

(a) A and B are exhaustive events
(b) A and B are independent events
(c) A and B are mutually exclusive events
(d) P(A|B) = .50 and P(B|A) = .2143
(e) P(A or B) = .85

Answer: (d)
Difficulty level: Easy

4.6.9: On a 100-mile stretch of the New Jersey Turnpike, four police officers are stationed apart to spot drunk drivers. All four police officers have somebody "pulled over" 60% of the time; thus, the chance of an officer being free and pulling over a drunk driver is 40%. If a drunk driver is driving that 100-mile stretch, what is the probability of him being arrested for drunk driving? Assume independence.

(a)   .1296
(b)   .8704
(c)   .6000
(d)   .4000
(e)   Cannot be determined

Answer: (b)
Difficulty level: Easy to Medium

4.6.10: Compute each of the following:

a.   5!
b.   10!/(8!2!)
c. $(.80)^3[(.20)^3]$
d. $6!/(3!3!)[(.80)^3(.20)^3]$

Answer: a.   120
          b.   45
          c.   .0041
          d.   .0820
Difficulty level: Easy to Medium

4.7.1: A local pizza place offers a choice of eleven pizza toppings. How many *unique* 3-item pizzas can possibly be made?

(a)   990
(b)   165
(c)   33
(d)   30
(e)   more than 10,000

Answer: (b)
Difficulty level: Easy

4.7.2: A comedy club has one show each night. In each show
exactly two comedians perform their comedy act. How many
consecutive nights can five comedians perform shows until a
show must have two of the same comedians of a previous show?

(a)   5
(b)   6
(c)   20
(d)   10
(e)   9

Answer: (d)
Difficulty level: Easy

4.7.3: A college professor must choose 5 students to help
him with a research project. 23 of the 50 students are
women. The probability the professor will choose 3 men and
2 women:

(a)   is less than 10%
(b)   is in between 10% and 25%
(c)   is in between 25% and 50%
(d)   is greater than 80%
(e)   cannot be determined

Answer: (c)
Difficulty level: Easy to Medium

4.7.4: A university professor must choose 5 students to help
him with a research project. 23 of the 50 students are
women. How many combinations of three men and two women are
there?

(a)   less than 1,000
(b)   at least 1,000 but less than 50,000
(c)   at least 500,000 but less than 1,000,000
(d)   greater than 1,000,000
(e)   at least 50,000 but less than 500,000

Answer: (c)
Difficulty level: Easy to Medium

Discrete Probability Distributions

5.1.1: A *random variable* is:

(a) a variable whose numerical value is known prior to the experiment taking place
(b) a variable whose numerical value is determined by the outcome of a random trial of an experiment
(c) a variable that represents the probability for an outcome of an experiment to take place
(d) a rule that assigns a name to each outcome of an experiment
(e) a variable that can only take on certain values over a given range of values

Answer: (b)
Difficulty level: Easy

5.1.2: Which of the following are examples of *continuous* random variables?

(a) the average distance of a discuss throw by olympic discuss final contestants
(b) the amount of unleaded gasoline consumed by a Honda Accord on a 1,200-mile trip to Colorado
(c) the average weight of a female bodybuilder competing in the Ms. Universe heavyweight division
(d) the average length of a human hand, measured from the top of the wrist to the outermost fingertip
(e) all of these

Answer: (e)
Difficulty level: Easy

5.1.3: Which of the following are examples of *discrete* random variables?

(a) the monthly volume of warehouse goods received by a warehouse from a goods supplier
(b) the diastolic blood pressure of a 60-year-old woman, measured to the nearest beat per minute
(c) the time required to complete a long distance race
(d) the first name of all teachers who teach introductory and intermediate courses in statistics
(e) more than one of these are discrete random variables

Answer: (b)
Difficulty level: Easy to Medium

5.1.4: A *discrete* random variable is one that:

(a)   can assume any value in a specified range of values
(b)   is known prior to the experiment taking place
(c)   is able to take on a countable number of values in an
      interval
(d)   can take on an infinite number of values in an interval
(e)   all of these

Answer: (c)
Difficulty level: Easy

5.1.5: Which of the following are examples of *continuous*
random variables?

(a)   the number of heads in three tosses of a coin
(b)   the average time required to iron a silk tie
(c)   the number of fours rolled in two rolls of a die
(d)   the number of pages in a mystery novel
(e)   none of these

Answer: (b)
Difficulty level: Easy

5.1.6: Which of the following are examples of *discrete*
random variables?

(a)   the total number of dots on an ordinary set of dice
(b)   the total number of doughnuts in a baker's dozen
(c)   the brand name of dishwasher units sold at Sears
(d)   the recording studio name used to record all of the
      songs on the album "So" by Peter Gabriel
(e)   none of these

Answer: (e)
Difficulty level: Easy to Medium

5.2.1: The following represents the probability distribution for the number of compact disks purchased by customers at a Discount Den record store on a given day.

X	P(x)
0	.23
1	.15
2	.29
3	.16
4	.09
5	.05
6	.03

a.   Find the mean or expected value.
b.   Find the variance.
c.   Find the standard deviation.

Answer: a.   2.00
          b.   2.52
          c.   1.59
Difficulty level: Easy

5.2.2: The probability distribution below shows the number of fire alarm calls received at a fire station in a large city during the 8 a.m. to 4 p.m. shift on weekdays.

Number of Fires	Probability
0	.20
1	.25
2	.20
3	.15
4	.10
5	.10

a.   Is this distribution continuous or discrete?
b.   What proportion of the time are there three or more fire alarm calls during the shift in question?
c.   Find the expected value of this probability distribution and write one sentence explaining its meaning.
d.   Find the variance of this distribution. What is the unit of measure for the variance?

Answer: a.   Discrete
          b.   .35
          c.   E(X) = 2.0. Over many, many days there will be an average of two alarms per day during the 8 a.m. to 4 p.m. shift.
          d.   Variance(X) = 2.5. Units on the variance are "fires squared."
Difficulty level: Easy to Medium

5.3.1: The following represents the probability distribution
for the number of bids submitted by a certified public
accounting firm prior to winning a competitive government
contract from the Small Business Administration.   The
expected value and standard deviation are:

x	1	2	3	4	5
P(X = x)	.1	.2	.4	.2	.1

(a)   3.0, 0.365
(b)   3.0, 1.20
(c)   3.0, 1.095
(d)   2.0, 1.20
(e)   2.0, 2.20

Answer: (b)
Difficulty level: Easy

5.3.2: The following represents the probability distribution
for the number of sales calls completed per day by a large
number of industrial salespeople with Great Lakes
Industries.   The expected value and standard deviation are:

x	0	1	2	3	4
P(X = x)	1/16	4/16	6/16	4/16	1/16

(a)   2.00, 2.00
(b)   2.00, 0.50
(c)   2.00, 1.00
(d)   32.0, 0.50
(e)   32.0, 1.00

Answer: (c)
Difficulty level: Easy

5.3.3: The following represents the probability distribution
for the number of cars rented per hour at the Rent-A-Car
counter at the Honolulu International Airport.   The expected
value and standard deviation are:

x	0	1	2	3	4
P(X = x)	.10	.25	.40	.20	.05

(a)   1.85, 1.03
(b)   1.85, 1.01
(c)   1.85, 0.55
(d)   1.95, 1.03
(e)   1.95, 1.01

Answer: (a)
Difficulty level: Easy

5.3.4: A physician has an office in each of two cities, approximately 30 miles apart. In City X, the expected number of patients seen in one day is 12, with a standard deviation of 4. In City Y, the expected number of patients seen by this physician is 10, with a standard deviation of 6. Assume that trials are independent.

a. What is the expected number of patients for the two cities?
b. What are the variance and standard deviation of the number of patients for the two cities?

Answer: a.   22
           b.   52, 7.21
Difficulty level: Easy to Medium

5.4.1: Two characteristics associated with the *binomial* distribution are:

(a)   two possible outcomes and continuous random variable
(b)   dependent trials and two possible outcomes
(c)   discrete random variable and dependent trials
(d)   two possible outcomes and discrete random variable
(e)   independent trials and continuous random variable

Answer: (d)
Difficulty level: Easy

5.4.2: Characteristics of the *binomial* distribution include:

(a)   the probability of a success or failure remains constant from one trial to the next
(b)   there being only two possible outcomes on each trial
(c)   trials being independent of one another
(d)   a discrete random variable
(e)   all of these

Answer: (e)
Difficulty level: Easy

5.4.3: The binomial and Poisson distributions *differ* in that:

(a)   only the binomial is based on a discrete random variable
(b)   only the Poisson has independent trials
(c)   only the binomial is appropriate for error processes
(d)   the parameters n and p determine the shape of the binomial distribution whereas the mean rate and interval length determine the shape of the Poisson distribution
(e)   the two distributions are identical

Answer: (d)
Difficulty level: Easy

5.4.4: Of the thousands of savings and loan institutions included in a liquidity study being made by the Federal Home Loan Bank Board, 50% are federally chartered and 50% are state chartered. An auditor randomly selects five savings and loans for an audit. What is the probability that <u>at least</u> 2 of the savings and loan institutions are federally chartered?

(a)   1/32
(b)   10/32
(c)   26/32
(d)   6/32
(e)   22/32

Answer: (c)
Difficulty level: Easy

5.4.5: Of the thousands of savings and loan institutions included in a liquidity study being made by the Federal Home Loan Bank Board, 50% are federally chartered and 50% are state chartered. An auditor randomly selects five savings and loans for an audit. What is the probability that exactly two of the savings and loan institutions are federally chartered?

(a)   1/32
(b)   10/32
(c)   26/32
(d)   6/32
(e)   22/32

Answer: (b)
Difficulty level: Easy

5.4.6: A shop foreman for Northern Manufacturing Corporation wants the supervisor to look at one of the defective parts that a new automatic drilling machine is producing. The machine has been producing defective parts at a rate of one every five parts. What is the probability that the shop foreman will find *fewer than* two defective parts in a randomly selected sample of four parts?

(a)   .4096
(b)   .8192
(c)   .5904
(d)   .1808
(e)   .8000

Answer: (b)
Difficulty level: Easy

5.4.7: Managers for the State Department of Transportation know that 70% of the people arriving at a toll plaza for a bridge have the exact or correct change. If 10 cars pass through the toll plaza in the next minute, what is the probability that *exactly* 6 of the cars will have exact or correct change?

(a)   210
(b)   .1176
(c)   .0081
(d)   .2001
(e)   .7999

Answer: (d)
Difficulty level: Easy

5.4.8: Find the following binomial probabilities:

a. $P(X = 2 \mid n = 15, p = .30)$
b. $P(X = 5 \mid n = 25, p = .50)$
c. $P(X = 9 \mid n = 10, p = .70)$
d. $P(X = 3 \mid n = 5, p = .30)$
e. $P(X \geq 4 \mid n = 5, p = .40)$
f. $P(X \leq 1 \mid n = 5, p = .05)$

Answer: a.   .0916
          b.   .0016
          c.   .1211
          d.   .1323
          e.   .0870
          f.   .9774
Difficulty level: Easy

5.4.9: If the process by which a certain manufacturer of automobile parts follows a binomial distribution with a known defective rate of 5%, what is the probability that *at least* one of a random sample of three auto parts will be defective?

(a)   .8574
(b)   .1426
(c)   .1354
(d)   approximately 1
(e)   approximately 0

Answer: (b)
Difficulty level: Easy

5.4.10: If the process by which a certain manufacturer of automobile parts follows a binomial distribution with a known defective rate of 5%, what is the probability that a random sample of three auto parts produces NO defectives?

(a)   .05
(b)   .95
(c)   .8574
(d)   .1286
(e)   .5000

Answer: (c)
Difficulty level: Easy

5.4.11: A manufacturer of electronic components knows that 10% of his components will be defective. What is the probability of at least one of a sample of 3 components is *not defective*?

(a)   .0010
(b)   .9703
(c)   .0297
(d)   .9000
(e)   .9990

Answer: (e)
Difficulty level: Easy

5.4.12: When p = .60 and n = 10, the shape of the binomial distribution for X, the number of successes, is:

(a)   skewed left with a mean of 6.0 and variance of 2.4
(b)   skewed right with a mean of 6.0 and variance of 2.4
(c)   skewed left with a mean of .60 and variance of .024
(d)   skewed right with a mean of .60 and variance of .024
(e)   symmetric with a mean of 6.0 and variance of 2.4

Answer: (a)
Difficulty level: Easy to Medium

5.4.13: When p = .40 and n = 10, the shape of the binomial distribution for X, the number of successes, is:

(a)   skewed left with a mean of 4.0 and variance of 2.40
(b)   skewed right with a mean of 4.0 and variance of 2.40
(c)   skewed left with a mean of .40 and variance of .024
(d)   skewed right with a mean of .40 and variance of .024
(e)   symmetric with a mean of 4.0 and variance of 2.40

Answer: (b)
Difficulty level: Easy to Medium

5.5.1: The *Poisson* distribution is characterized by:

(a)   discrete outcomes
(b)   sampling over some continuous space
(c)   events occurring independently
(d)   the probability of occurrence changing proportionately
      per unit of space
(e)   all of these

Answer: (e)
Difficulty level: Easy to Medium

5.5.2: The defects in the rubber covering on a particular type of telephone cable follow a *Poisson* distribution with a mean rate of .00065 per lineal foot. The probability of *fewer than* 3 defects in a spool of 10,000 lineal feet is:

(a)   .0687
(b)   .9313
(c)   .0431
(d)   6.50
(e)   Cannot be determined

Answer: (c)
Difficulty level: Easy

5.5.3: The defects in the rubber covering on a particular type of telephone cable follow a *Poisson* distribution with a mean rate of .00065 per lineal foot. The probability of exactly two defects in a spool of 10,000 lineal feet is:

(a) .0113
(b) .9887
(c) .0318
(d) 6.50
(e) Cannot be determined

Answer: (c)
Difficulty level: Easy

5.5.4: The errors per page made on a typing test follow a *Poisson* distribution with a mean rate of 4.3 errors per page.

a. Calculate the probability that a single page of typing contains no errors.
b. Calculate the probability that 2 pages of typing will contain no more than 2 errors.

Answer: a.   .0136
          b.   .0086
Difficulty level: Easy to Medium

5.5.5: At a local bank, the average number of customer arrivals per hour is believed to follow a *Poisson* distribution with a mean rate of 4.5 customers per hour.

a. What is the probability that 6 or more customers arrive in a given hour?
b. What is the probability that less than 3 customers arrive in a given hour?
c. If the bank has 2 tellers that can serve 4 customers each per hour, what is the probability that there is a need for a third teller on a given hour?

Answer: a.   .297
          b.   .174
          c.   .040
Difficulty level: Easy to Medium

5.6.1: Two probability distributions that are based on *discrete* random variables are:

(a) the uniform and normal distributions
(b) the Poisson and normal distributions
(c) the uniform and normal distribution
(d) the Poisson and binomial distributions
(e) the hypergeometric and normal distributions

Answer: (d)
Difficulty level: Easy

5.7.1: In a group of eight computer programmers, there are five who have training in the computer language required on the next programming job. If three programmers are selected at random, what is the probability that exactly two will have the necessary training in the computer language?

(a) 5/28
(b) 15/28
(c) 2/3
(d) 10/24
(e) approximately zero

Answer: (b)
Difficulty level: Easy to Medium

5.8.1: The expected value of the *sum* of two discrete random variables:

(a) is equal to the sum of their expected values
(b) is equal to the average of their expected values
(c) is equal to the larger expected value divided by the smaller expected value
(d) is equal to the larger expected value minus the smaller expected value
(e) is equal to the product of their expected values

Answer: (a)
Difficulty level: Easy

5.8.2: The standard deviation of the *difference between* two independent discrete variables is equal to:

(a)   the sum of the individual standard deviations
(b)   the square root of the sum of the individual variances
(c)   the sum of the square root of the individual variances
(d)   the average of the two individual standard deviations
(e)   the larger standard deviation minus the smaller standard deviation

Answer: (b)
Difficulty level: Easy to Medium

5.8.3: A discrete random variable U is defined by the linear function: U = AX + B, where A and B are constants. The expected values of these two random variables, U and X:

(a)   are identical for all positive values of A and B
(b)   are identical for all values of B given that A = 1
(c)   satisfy the equation $E(U) = AE(X)$
(d)   satisfy the equation $E(U) = AE(X) + B$
(e)   satisfy the equation $E(U) = A^2E(X) + B$

Answer: (d)
Difficulty level: Easy

5.8.4: A discrete random variable U is defined by the linear function: U = AX + B, where A and B are constants. The variances of these two random variables, U and X:

(a)   are identical for all positive values of A and B
(b)   are identical for all values of B given that A = 1
(c)   satisfy the equation $V(U) = AV(X)$
(d)   satisfy the equation $V(U) = A^2V(X) + B$
(e)   none of these

Answer: (b)
Difficulty level: Medium

Continuous Probability Distributions

6.1.1: A probability distribution for a continuous random variable is specified by:

(a) a discrete function
(b) a probability density curve
(c) a spike chart
(d) the normal approximation to the binomial
(e) the normal approximation with continuity correction

Answer: (b)
Difficulty level: Easy

6.1.2: Which of the following statements about the normal distribution is most correct?

(a) The normal distribution is the least commonly used and has the fewest business applications of all continuous probability distributions.
(b) The normal distribution is the most commonly used yet has the fewest business applications of all continuous probability distributions.
(c) The normal distribution is the least commonly used yet has the most business applications of all continuous probability distributions.
(d) The normal distribution is the most commonly used and has the most business applications of all (discrete and continuous) probability distributions.
(e) None of these.

Answer: (d)
Difficulty level: Easy

6.1.3: A property of *continuous* probability distributions such as the normal distribution is that:

(a) as with discrete random variables, the probability distribution can be approximated by a smooth curve.
(b) as with discrete random variables, endpoints make a difference in probability calculations.
(c) unlike discrete random variables, probabilities can be found using tables.
(d) unlike discrete random variables, the probability that a continuous random variable equals a specific value is zero; hence $P(X = a) = 0$.
(e) probabilities for continuous random variables can be approximated using discrete techniques such as the continuity correction.

Answer: (d)
Difficulty level: Easy

6.2.1: The mean and variance for a *uniform* distribution (with lower and upper endpoints of a and b) are given by:

(a) $1/(b - a)$ and $1/(b - a)^2$, respectively
(b) $(1/2)(a + b)$ and $(1/2)^2(a + b)^2$, respectively
(c) $(1/2)(a + b)$ and $(1/12)(a + b)^2$, respectively
(d) $(1/2)(a + b)$ and $(1/12)(b - a)^2$, respectively
(e)   none of these

Answer: (d)
Difficulty level: Easy

6.2.2: A data analysis system generates random numbers according to the *uniform* distribution, which has a probability density function given by: $f(X) = 1/(b - a)$, where $0 \leq X \leq 2$. Compute or find the following:

a. $P(0 \leq X \leq .5)$
b. $P(0 \leq X \leq 1.2)$

Answer: a.   .25
          b.   .60
Difficulty level: Easy

6.2.3: A study shows that employees who begin their work day at 8:00 a.m. vary their times of arrival *uniformly* over the range 7:40 a.m. to 8:30 a.m. The probability that a random employee arrives between 8:00 a.m. to 8:10 a.m. is:

(a)   40%
(b)   20%
(c)   10%
(d)   approximately 50%
(e)   approximately 16.7%

Answer: (b)
Difficulty level: Easy

6.3.1: A *normal* distribution is characterized by:

(a)   a mean of $(1/2)(b + a)$ and a variance of $(1/12)(b - a)^2$
(b)   a mean of p and a variance of $np(1-p)$, provided $np \geq 5$
(c)   a mean of μ and a standard deviation of σ/n
(d)   a mean of μ and a standard deviation of σ
(e)   a mean of 0 and a variance of 1

Answer: (d)
Difficulty level: Easy

6.3.2: The *standard normal* distribution is characterized by:

(a) a mean of 0 and a variance of 1
(b) a mean of $(1/2)(b + a)$ and a variance of $(1/12)(b - a)^2$
(c) a mean of p and a variance of $np(1-p)$, provided $np \geq 5$
(d) a mean of μ and a standard deviation of σ
(e) a mean of μ and a standard deviation of σ/n

Answer: (a)
Difficulty level: Easy

6.3.3: An important characteristic of the *normal* curve is that it may be totally described by the:

(a) probability of a success p and number of trials n
(b) variance and standard deviation
(c) mean and standard deviation
(d) mean, median, and mode
(e) mean rate

Answer: (c)
Difficulty level: Easy

6.3.4: Assume that the GMAT scores of students planning to enter graduate school in business are *normally* distributed with a mean of 450 and a standard deviation of 50.

a. What is the probability that a randomly chosen applicant has a score in between 525 and 575? What proportion of students score in between 400 and 500?
b. If a certain university has a minimum GMAT score of 550 and 50 students apply to their MBA program, how many of the students will meet the minimum standard?

Answer: a. .0606, .6826
        b. only 1 student
Difficulty level: Easy to Medium

6.3.5: Given that Z is the *standard normal* random variable, what is $P(.53 \leq Z \leq 2.42)$?

(a) .4922
(b) .2019
(c) .2903
(d) .9922
(e) .7903

Answer: (c)
Difficulty level: Easy

6.3.6: Given that Z is the *standard normal* random variable, what is P(0 ≤ Z ≤ 1.63)?

(a) .0516
(b) .4484
(c) .9484
(d) .8968
(e) .3764

Answer: (b)
Difficulty level: Easy

6.3.7: Given that Z is the *standard normal* random variable, what is P(-1.63 ≤ Z ≤ 1.63)?

(a) .0516
(b) .4484
(c) .9484
(d) .8968
(e) .1032

Answer: (d)
Difficulty level: Easy

6.3.8: Given that Z is the *standard normal* random variable, what is P(-1.63 ≤ Z ≤ 0.53)?

(a) -.2465
(b) .2465
(c) .3497
(d) .7465
(e) .6503

Answer: (e)
Difficulty level: Easy

6.3.9: Given that Z is the *standard normal* random variable, what is P(-1.68 ≤ Z ≤ -0.29)?

(a) -.3394
(b) .3394
(c) .4324
(d) .6606
(e) .5676

Answer: (b)
Difficulty level: Easy

6.3.10: Given that X is a *normal* random variable with a mean of 500 and a variance of 400, what is P(475 ≤ X ≤ 550)?

(a) .8882
(b) .4938
(c) .3944
(d) .1118
(e) .0756

Answer: (a)
Difficulty level: Easy

6.3.11: Given that X is a *normal* random variable with a mean of 500 and a variance of 100, what is P(475 ≤ X ≤ 500)?

(a) .4938
(b) .9938
(c) .0062
(d) .5062
(e) .2802

Answer: (a)
Difficulty level: Easy

6.3.12: Given that X is a *normal* random variable with a mean of 500 and a variance of 100, what is P(500 ≤ X ≤ 550)?

(a) .1915
(b) .6915
(c) .3005
(d) approximately 50%
(e) approximately 100%

Answer: (d)
Difficulty level: Easy

6.3.13: Given that X is a *normal* random variable with a mean of 100 and a variance of 100, what is P(75 ≤ X ≤ 100)?

(a) .4938
(b) .9938
(c) .0062
(d) .5062
(e) .2802

Answer: (a)
Difficulty level: Easy

6.3.14: Given that X is a *normal* random variable with a mean of 100 and a variance of 100, what is $P(80 \leq X \leq 125)$?

(a)   .0166
(b)   .4710
(c)   .0290
(d)   .9710
(e)   .9834

Answer: (d)
Difficulty level: Easy

6.3.15: Given that X is a *normal* random variable with a mean of 50 and a variance of 100, what is $P(45 \leq X \leq 50)$?

(a)   .1915
(b)   .6915
(c)   .3005
(d)   approximately 50%
(e)   approximately zero

Answer: (a)
Difficulty level: Easy

6.3.16: Given that X is a *normal* random variable, the probability that a given value of X is below its mean is:

(a)   approximately 0
(b)   approximately 1
(c)   approximately 50%
(d)   unknown, since the distribution might be skewed
(e)   unknown, since the mean and variance are unknown

Answer: (c)
Difficulty level: Easy to Medium

6.3.17: The relation between the standard normal random variable Z and a normal random variable X is that:

(a)   only the normal random variable X is continuous
(b)   only the standard normal random variable is continuous
(c)   the standard normal random variable Z counts the number of standard deviations that the value of a normal random variable X is away from its mean
(d)   the standard normal random variable cannot be negative
(e)   values of a normal random variable cannot be negative

Answer: (c)
Difficulty level: Easy

CHAPTER 6   **71**

6.4.1: Sixty percent of the applicants for membership in an
insurance company's "Preferred Risk Club" are accepted. The
remaining forty percent of the applicants are not accepted.
Applicants who are accepted receive a discount of auto
insurance premiums and a semi-annual newsletter containing
safe-driving tips. The probability that *at most* 195 out of
300 applicants are accepted into the "Preferred Risk Club":

(a)   is 1.83
(b)   is .4664
(c)   is .0336
(d)   is .9664
(e)   is .9616

Answer: (d)
Difficulty level: Medium

6.4.2: If 300 registered voters are surveyed and the
proportion who favor an increase in sales tax is 54%, what
is the probability that the proportion who favor the tax
increase is at least 49% but no greater than 58%?

(a)   .1094
(b)   .8906
(c)   .8768
(d)   .1232
(e)   .5000

Answer: (b)
Difficulty level: Medium

6.4.3: Electronics Corporation of America, a company with a
very large number of employees, has approximately 44% women
employees. A group of 50 employees is randomly selected to
serve on a grievance review board. Compute the following.

a.   What is the probability that at least 30 women will be
     selected to be on the grievance review board?
b.   What is the probability that at most 25 women will be
     selected to be on the grievance review board?
c.   What is the probability that at the number of women
     selected to be on the grievance board will be at least
     20 but no greater than 30?

Answer: a.   .0778
          b.   .1587
          c.   .8335
Difficulty level: Easy to Medium

6.4.4: A *continuity correction* is an adjustment that is made:

(a)   anytime the normal distribution is used
(b)   when using the binomial distribution whenever np < 5
(d)   anytime we are dealing with a discrete distribution
(d)   when using a continuous probability distribution to approximate a discrete probability distribution
(e)   when using a discrete probability distribution to approximate a continuous probability distribution

Answer: (d)
Difficulty level: Easy

6.4.5: The *continuity correction* is used to:

(a)   correct a mistake in the calculations
(b)   improve the accuracy of a probability approximation
(c)   solve a uniform probability problem
(d)   solve a normal probability problem
(e)   all of these

Answer: (b)
Difficulty level: Easy

6.4.6: Forty-five percent of all persons employed by the Southeast Manufacturing Company are college graduates. A random sample of 100 of these employees is selected. Assuming that trials are independent, what is the probability that at least 55 of the employees sampled are college graduates?

(a)   .4826
(b)   .9826
(c)   .0174
(d)   .9719
(e)   .0281

Answer: (c)
Difficulty level: Easy to Medium

6.5.1: The time that a tourist waits to board a trolley car at Polk Street in San Francisco has a *uniform* distribution with a mean of 7 minutes and a standard deviation of 3 minutes. The probability that a randomly selected person waits between 5 and 7 minutes is:

(a)   0
(b)   about 20%
(c)   in between 25% and 50%
(d)   greater than 50%
(e)   unable to be determined with the information given

Answer: (b)
Difficulty level: Challenging

6.5.2: A wooden beam produced by Carolina Atlantic Corporation is normally distributed with a mean breaking strength of 1500 pounds and a standard deviation of 100 pounds. What is the percentage of all such beams that have a breaking strength in between 1450 and 1600 pounds?

(a)   .1915
(b)   .3413
(c)   .5328
(d)   .4628
(e)   approximately 75%

Answer: (c)
Difficulty level: Easy

6.5.3: The service lives of electron tubes produced by Tensor Corporation are normally distributed; 93.94% of the tubes have lives greater than 2,000 hours, and 7.35% have lives greater than 17,000 hours. What are the mean and the standard deviation of service lives for the electron tubes?

(a)   8,856 hours and 4,650 hours, respectively
(b)   5,000 hours and 9,750 hours, respectively
(c)   9,750 hours and 5,000 hours, respectively
(d)   9,750 hours and 500 hours, respectively
(e)   cannot be determined

Answer: (c)
Difficulty level: Challenging

6.5.4: The resistances of carbon resistors of 1300 ohms nominal value are normally distributed with μ = 1300 ohms and σ = 200 ohms. Compute or determine the following.

a.   What proportion of these resistors will have resistances greater than 1000 ohms?
b.   What proportion of these resistors will have resistances that differ from the mean by *less than 2%* of the mean on either side?
c.   What proportion of these resistors will have resistances that differ from the mean by *less than 4%* of the mean on either side?

Answer: a.   .0668
           b.   .1034
           c.   .2052
Difficulty level: Medium

6.5.5: Atlantic Bank receives 42% of its bank card applications from unmarried people. What is the probability that in the next 150 applications 70 or fewer applicants will be unmarried? (Hint: Remember the continuity correction.)

(a)   .1075
(b)   .3925
(c)   .8925
(d)   .6075
(e)   cannot be determined

Answer: (c)
Difficulty level: Easy to Medium

6.5.6: An Auto Parts store has found that 4% of the spark plugs received from a supplier are not properly gapped. A shipment of 1000 spark plugs just arrived from the supplier.

a.   What is the expected number of spark plugs that are not properly gapped?
b.   What is the probability that there will be at most 35 spark plugs in the shipment that are not properly gapped? Use the continuity correction.
c.   What is the probability that there will be at most 50 spark plugs in the shipment that are not properly gapped? Use the continuity correction.

Answer: a.   40
           b.   .1867
           c.   .9545
Difficulty level: Easy to Medium

6.6.1: The *exponential* probability distribution may be used to find which of the following probabilities?

(a) the proportion of customers arriving at a bank teller's window will approach the window within three minutes of the previous customer
(b) the proportion of long-distance telephone calls received at a switchboard that are completed within five minutes
(c) the proportion of new automobiles that will have transmission failures within the first 25,000 miles
(d) the proportion of machines that will break down during an eight-hour night shift
(e) all of these

Answer: (e)
Difficulty level: Easy

6.6.2: The number of customers arriving at a bank teller's window at Commercial Bank follows the *Poisson* distribution with a mean rate of .75 customers per minute. If the time between arrivals is less than or equal to three minutes, then the teller can provide banking services without irritating customers owing to an annoying waiting time. Compute or find the following.

a. The mean and standard deviation of the time in between customer arrivals at the bank teller's window.
b. The proportion of customers for whom the bank teller provides service <u>without</u> an annoying waiting time.

Answer: a.   1.33 minutes/customer, 1.33 minutes/customer.
           b.   .8946.
Difficulty level: Easy to Medium

6.6.3: Environmental concerns in Washington County have resulted in a hazardous materials clean-up team organized by the Department of Public Safety. The number of hazardous chemical spills in the county that require the intervention of the clean-up team follows a *Poisson* distribution with a mean rate of 2.0 interventions per month. Compute or find the following.

a.   The mean and standard deviation of the time in between interventions of the clean-up team for hazardous chemical spills.
b.   The proportion of interventions that occur within one month of a previous intervention.
c.   The proportion of interventions that occur within two months of a previous intervention.

Answer: a.   .50 months/intervention
          b.   .8647
          c.   .9817
Difficulty level: Easy to Medium

Sampling Distributions

7.1.1: *Random sampling* is best defined as a method of sampling:

(a)   that chooses items in such a way that each item and each group of n items is equally likely to be chosen or included in the sample.
(b)   that chooses items in a systematic way that necessarily includes some items and does not include other items.
(c)   that sorts items into clusters and then samples from these clusters in a random way.
(d)   that sorts items into strata and then samples from these strata in a way representative of the population.
(e)   in which the elements are ordered in an array and then every 10th observation is sampled.

Answer: (a)
Difficulty level: Easy

7.1.2: The importance of sampling distributions arises from the fact that:

(a)   often we are interested in describing facts about a population and must do so by relying on sample information obtained from a random sample.
(b)   often we are interested in describing facts about a population and must do so by describing how the values of the population parameter varies across samples.
(c)   often we are interested in describing facts about a population and must do so by describing how the values of the sample statistic vary across samples.
(d)   the sampling distribution of any sample statistic is always normally distributed, regardless of whether the population is normally distributed.
(e)   the Central limit theorem only applies when we know that the population is normally distributed.

Answer: (c)
Difficulty level: Easy

7.1.3: The probability distribution of a statistic, such as a proportion or a mean, is known as:

(a)   the mean and variance of the statistic
(b)   the infinite universe
(c)   the underlying parameters
(d)   the sampling distribution of the statistic
(e)   the set of all possible random samples

Answer: (d)
Difficulty level: Easy

7.1.4: We take random samples from populations to study or estimate facts about the populations. These facts are usually expressed in terms of numbers called:

(a)   statistics or parameters, depending on whether a random sampling procedure was used.
(b)   statistics or parameters, depending on whether the sample size is sufficiently large.
(c)   statistics or parameters, depending on whether we are trying to determine an unknown mean or proportion.
(d)   parameters.
(e)   statistics.

Answer: (e)
Difficulty level: Easy

7.1.6: The sampling distribution of the sample mean:

(a)   is only well defined for finite populations.
(b)   is only well defined for normal populations.
(c)   more than one of these statements is correct.
(d)   is a normal distribution for large sample sizes.
(e)   is always symmetric so that mean = median = mode.

Answer: (d)
Difficulty level: Easy to Medium

7.2.1: The expected value of the distribution of sample means will equal the population mean:

(a)   only for populations with a known standard deviation.
(b)   only when random sampling procedures are used.
(c)   only when sampling is done with replacement.
(d)   only for normal populations.
(e)   for all populations.

Answer: (e)
Difficulty level: Easy

7.2.2: If the population is *infinite*, then the variance of the sample mean is equal to:

(a) σ^2
(b) σ^2/n
(c) $\sigma^2/[$the square root of n$]$
(d) $[\sigma^2/n][(N - n)/(N - 1)]$
(e) $[\sigma^2/n][$the square root of $(N - n)/(N - 1)]$

Answer: (b)
Difficulty level: Easy

7.2.3: The term $[(N - n)/(N - 1)]$ is used:

(a)   only when random sampling procedures are used.
(b)   when the population is infinite and sampling is done with replacement.
(c)   when the statistician does not know the true value of the population standard deviation.
(d)   when the population is infinite and sampling is done without replacement.
(e)   when the population is finite and sampling is done without replacement.

Answer: (e)
Difficulty level: Easy to Medium

7.2.4: If sampling is done *without replacement* from a finite population, the variance of the sample mean is equal to:

(a) σ^2
(b) σ^2/n
(c) $\sigma^2/[$the square root of n$]$
(d) $[\sigma^2/n][(N - n)/(N - 1)]$
(e) $[\sigma^2/n][$the square root of $(N - n)/(N - 1)]$

Answer: (d)
Difficulty level: Easy

7.2.5: The *finite population correction factor*:

(a) is always negative.
(b) is always greater than 1.
(c) is always less than or equal to 1.
(d) is used whenever random sampling procedures are used.
(e) more than one of these is correct.

Answer: (c)
Difficulty level: Medium

7.2.6: The finite population correction factor MUST be used:

(a) whenever random sampling procedures are used.
(b) whenever sampling is done with replacement.
(c) whenever the statistician knows the standard deviation.
(d) whenever sampling without replacement from a finite
     population.
(e) more than one of these statements is correct.

Answer: (d)
Difficulty level: Easy to Medium

7.2.7: A simple random sample of size n = 36 is to be
selected from a population with σ = 18. Compute the value
of the standard error of the sample mean in each case.

a.   The population is infinite.
b.   The population size is N = 50.
c.   The population size is N = 500.

Answer: a.   3.00
           b.   1.60
           c.   2.89
Difficulty level: Easy to Medium

7.2.8: A sample of size n = 100 is drawn from an infinite
population with mean μ = 50 and standard deviation σ = 20.
The expected value and standard deviation of the sampling
distribution of the sample mean:

(a) are 50 and 20, respectively
(b) are 50 and 2, respectively
(c) are 50 and .20, respectively
(d) are unknown and 2, respectively
(e) cannot be determined without further information

Answer: (b)
Difficulty level: Easy to Medium

7.2.9: A sample of size n = 100 is drawn from a finite population with mean μ = 50 and standard deviation σ = 20. The expected value and standard deviation of the sampling distribution of the sample mean:

(a)   are 50 and 20, respectively.
(b)   are 50 and 2, respectively.
(c)   are 50 and .20, respectively.
(d)   are unknown and .20, respectively.
(e)   cannot be determined without further information.

Answer: (e)
Difficulty level: Medium to Challenging

7.2.10: A sample of size n = 25 is drawn from an infinite population with mean μ = 50 and standard deviation σ = 20. The expected value and standard deviation of the sampling distribution of the sample mean:

(a)   are 50 and 4, respectively.
(b)   are 50 and 20, respectively.
(c)   are 50 and .80, respectively.
(d)   cannot be determined without further information.
(e)   are both unknown since the sample size is less than 30.

Answer: (a)
Difficulty level: Easy to Medium

7.3.1: If a population is known to be *normally distributed*, then it follows from this that:

(a)   the sample mean must equal the population mean.
(b)   the expected value of the sample mean must equal μ.
(c)   the sample standard deviation must equal σ.
(d)   the sampling distribution must be normally distributed.
(e)   none of these.

Answer: (d)
Difficulty level: Medium

7.3.2: If a population is *not* normally distributed, then:

(a)   the sample mean must equal the population mean.
(b)   the sample variance must equal the population variance.
(c)   the sampling distribution is normally distributed.
(d)   the sampling distribution is normally distributed if
      and only if the sample size is large.
(e)   the sampling distribution will not be normally
      distributed, even if the sample size is very large.

Answer: (d)
Difficulty level: Easy to Medium

7.4.1: The lengths of individual machine parts coming off a
production line at Morton Metalworks are *normally
distributed* around a mean of μ = 30.00 centimeters with a
standard deviation of σ = .10 centimeters. An inspector
just took a random sample of 4 parts and obtained a sample
mean of 29.875 centimeters. What is the probability of
getting a sample mean this short or shorter?

(a)   .8944
(b)   .9938
(c)   .1056
(d)   .0062
(e)   cannot be determined

Answer: (d)
Difficulty level: Easy

7.4.2: The increase in the yield of wheat (in bushels) when
a pesticide is used on a plot of ground at California
Agronomics Land & Farming, Inc., is *normally distributed*
with a mean of μ = 50 and a variance of σ = 100. What is
the probability that the mean increase in wheat yield for a
sample of n = 25 plots will differ from the population mean
by *less than* 4 bushels?

(a)   .4772
(b)   .9544
(c)   .0456
(d)   .1586
(e)   It cannot be determined since n < 30

Answer: (b)
Difficulty level: Medium

7.4.3: Weekly receipts at the Star Restaurant are *normally distributed* with a mean of μ = $17,000 and a standard deviation of σ = $2000. The probability that in a sample of n = 10 weeks the average receipts are *less than* the restaurant's break-even point of $15,000 is:

(a) .0008
(b) .9992
(c) .4992
(d) .3413
(e) .1587

Answer: (a)
Difficulty level: Easy

7.4.4: The time to complete a task involving a production process is *normally distributed* with a mean μ = 13 minutes and a standard deviation of σ = .60 minutes. In a sample of 36 completion times, what is the probability that the average completion time is at most 13.25?

(a) .4938
(b) .0062
(c) .9938
(d) at most 16%
(e) at least 84%

Answer: (c)
Difficulty level: Easy to Medium

7.5.1: As a result of the Central Limit Theorem:

(a) we need not take random samples, just large ones.
(b) using the normal approximation to the binomial is always justified.
(c) we can use small sample sizes provided we are sampling from normal populations.
(d) we may use normal theory for inferences about the sample mean regardless of the sampling procedures used.
(e) we may use normal theory for inferences about the population mean regardless of the form of the population if and only if the sample size is large enough.

Answer: (e)
Difficulty level: Easy to Medium

7.5.2: The *Central limit theorem* states that:

(a)   the sampling distribution of the sample mean will be
      approximately normally distributed provided that the
      sample size n is sufficiently large.
(b)   the sampling distribution of the sample proportion will
      be approximately normally distributed provided the
      population is normally distributed.
(c)   the sampling distribution of the sample mean will be
      approximately normally distributed provided the
      population is normally distributed.
(d)   if the sample size n is large, then Z follows a
      standard normal distribution, provided that the
      population is normally distributed.
(e)   more than one of these is correct.

Answer: (a)
Difficulty level: Easy to Medium

7.5.3: The relationship between the shape of the population
distribution and the shape of the sampling distribution
of the mean is *best* explained by:

(a)   the standardized sample mean or Z variable
(b)   sampling from a finite population
(c)   the central limit theorem
(d)   the population finite correction factor
(e)   simple random sampling

Answer: (c)
Difficulty level: Easy

7.5.4: Using the normal approximation to the binomial is
justified whenever:

(a)   N is finite
(b)   N is infinite
(c) $np(1 - p) \geq 5$
(d) $n < 30$
(e) $p = .50$ and $np(1 - p) < 5$

Answer: (c)
Difficulty level: Easy to Medium

7.5.5: The daily catch of a small tuna-fishing fleet from Ocean Foods Company averages μ = 130 tons with a standard deviation of σ = 42 tons. The probability that during a sample of n = 36 fishing days the total weight of the catch will be at least 3520 tons is:

(a) .4326
(b) .9326
(c) .0674
(d) at most 48.9%
(e) cannot be determined unless the population is normal

Answer: (b)
Difficulty level: Easy to Medium

7.5.6: The probability of six-month incomes of account executives has a mean of μ = $20,000 and a standard deviation of σ = $5,000. Compute or answer the following.

a.   A single account executive is selected at random and her six-month income is $20,000. Can it be said that her six-month income exceeds 50% of all account executives?
b.   A random sample of only n = 16 account executives is taken. What is the probability that the sample mean six-month income is less than $20,500?
c.   A random sample of n = 64 account executives is taken. What is the probability that the sample mean six-month income is in between $20,500 and $22,000?

Answer: a.   No. If the distribution were skewed left, for example, then her income would fall below the median income.
          b.   Cannot be determined (since n < 30).
          c.   .4993 - .2881 = .2112.
Difficulty level: Easy to Medium

7.5.7: Seventy-five percent of newly organized businesses experience cash flow problems during their first year of operation. A consultant for the Small Business Administration takes a random sample of n = 50 small businesses that have been in operation for one year. Compute or answer the following.

a.   What are the mean and variance for the *fraction* of new small businesses that experience cash flow problems during the first year of operation?
b.   What is the probability that the sample proportion of the newly organized businesses that are experiencing cash flow problems is at least 80%?

Answer: a. $E(\hat{p}) = p = .75$, $\sigma^2(\hat{p}) = p(1 - p)/n = .0038$
          b. $.5000 - .2939 = .2661$
Difficulty level: Easy to Medium

7.5.8: A sample of size n = 169 is drawn from a *skewed* infinite population with mean μ = 50 and standard deviation σ = 20. What is the probability that the sample mean is at least 54?

(a)   .4953
(b)   .5000
(c)   .9953
(d)   .0047
(e)   cannot be determined since the population is skewed

Answer: (d)
Difficulty level: Easy to Medium

7.5.9: Seventy-five percent of a school's law class passes the state bar examination on the first attempt. If a randomly selected group of 60 of this school's law graduates take the examination, what is the probability that *at least* 80% will pass the examination on the first attempt?

(a)   .3133
(b)   .8133
(c)   .1867
(d)   .6867
(e)   Cannot be determined since the population is skewed

Answer: (c)
Difficulty level: Easy

7.5.10: Time lost at work due to employee absenteeism is an important problem for many companies. The human resources department of Western Electronics has studied the distribution of time lost due to absenteeism by individual employees. During a one-year period the department found a mean of μ = 21 days and a standard deviation of σ = 10 days. A group of 49 employees is selected at random to participate in a program that allows a flexible work schedule, which the human resources department hopes will decrease the amount of employee absenteeism in the future. Answer the following.

a. What is the probability that the sample mean value for time lost due to absenteeism for this group of employees will exceed 21 days? <u>Explain</u>.
b. What is the probability that the sample mean value for time lost due to absenteeism for this group of employees will be in between 19 and 23 days?
c. What values that are an equal distance from the mean amount of time lost due to absenteeism encompass 80% of all possible sample mean values?
d. What values that are an equal distance from the mean amount of time lost due to absenteeism encompass 95% of all possible sample mean values?
e. What values that are an equal distance from the mean amount of time lost due to absenteeism encompass 98% of all possible sample mean values?

Answer: a. .5000, since the normal distribution is symmetric and bell-shaped
         b. .8384
         c. [19.17 to 22.83 days absent]
         d. [18.20 to 23.80 days absent]
         e. [18.67 to 24.33 days absent]
Difficulty level: Easy to Challenging

7.6.1: A sample of size n = 16 is drawn from a *normally distributed* population with mean μ = 20 and standard deviation σ = 8. The probability that the sample mean is greater than 15 is:

(a) .4938
(b) .9938
(c) .0062
(d) .5000
(e) cannot be determined since n is small and the population is skewed

Answer: (b)
Difficulty level: Easy

7.6.2: Digital industries uses statistical quality control
to monitor the production process during the manufacture of
memory chips. The electrical usage across a conducting zone
for a memory chip is *normally distributed* with a mean of
10.05 microvolts and a standard deviation of .92 microvolts.
A quality control manager randomly samples 4 memory chips.
What is the probability that the sample mean is in between
8.67 and 11.43 microvolts?

(a)   .0026
(b)   .9974
(c)   .4987
(d)   .0013
(e)   Unknown, since n < 30.

Answer: (b)
Difficulty level: Easy to Medium

7.6.3: A type of cathode ray produced at Amp, Inc., has a
mean life of 10,000 hours and a known variance of 14,400
(squared hours). A few tubes have extraordinary long lives,
so the distribution is slightly skewed to the right. If we
take a large number of random samples of n = 36 tubes each
and compute the mean life for each, between what limits
would 50% of the sample means be expected to lie (symmetric
around the mean)?

(a)   10,000 ± (120/6)
(b)   10,000 ± (3.89)(120/6)
(c)   10,000 ± (0.67)(120/6)
(d)   10,000 ± (3.89)(14,400/6)
(e)   10,000 ± (0.67)(14,400/6)

Answer: (c)
Difficulty level: Medium

7.6.4: The population of times required to perform an
assembly line task is skewed with a mean of μ = 3 minutes
and a standard deviation of σ = .2 minutes. If we take a
large number of samples of 36 employees each, between what
limits would 95% of the sample means be expected to lie?

(a)   3 ± (.2/6)
(b)   3 ± (1.645)(.2/36)
(c)   3 ± (1.645)(.2/6)
(d)   3 ± (1.96)(.2/36)
(e)   3 ± (1.96)(.2/6)

Answer: (d)
Difficulty level: Medium

7.7.1: If X and Y are independent random variables, then which of the equations below must be true?

(a) Variance(X - Y) = Variance(X) - Variance(Y)
(b) Variance(X - Y) = Variance(X) + Variance(Y)
(c) Variance(X - Y) = [Variance(X)][Variance(Y)]
(d) Variance(X - Y) = [Variance(X)]/[Variance(Y)]
(e) All of these

Answer: (b)
Difficulty level: Easy to Medium

7.7.2: If X and Y are random variables with a __negative__ covariance, which of the following *must* be true?

(a) Variance(X + Y) = Variance(X) + Variance(Y)
(b) Variance(X - Y) = Variance(X) - Variance(Y)
(c) Variance(X - Y) < Variance(X + Y)
(d) Variance(X - Y) > Variance(X + Y)
(e) More than one of these are true

Answer: (d)
Difficulty level: Medium

7.7.3: Yields for municipal bonds (X_1) over a period have had a mean of $\mu_1 = .10$ and a standard deviation of $\sigma_1 = .02$. Yields for industrial bonds (X_2) have had a mean of $\mu_2 = .13$ and a standard deviation of $\sigma_2 = .03$. Independent random samples, each of size n = 49, are obtained from the two (very large) populations of bonds. Answer the following.

a.   What is the expected value for the difference in sample mean yields?
b.   What is the standard deviation of the difference in sample mean yields?
c.   What is the probability that the difference in sample mean yields, will be less than -.02?
d.   Is there strong evidence that the average performance of industrial bonds exceeds that of municipal bonds?

Answer: a. $E(\overline{X}_1 - \overline{X}_2) = -.03$

           b.   Standard deviation$(\overline{X}_1 - \overline{X}_2) = .0052$

           c.   .5000 + .4738 = .9738

           d.   Yes, since $P[(\overline{X}_1 - \overline{X}_2) < 0]$ is approximately 1
Difficulty level: Easy to Medium

7.7.4: A manager is interested in trying a new selling technique to see if it increases sales. He used the new technique in store 1 and stayed with the old technique in store 2. Daily sales for 150 days were recorded for both stores. Store 1 daily sales averaged $2,000 with a standard deviation of $160. Store 2 daily sales averaged $1,960 with a standard deviation of $140. What is the probability that the sample mean for store 2 is less than or equal to the sample mean for store 1?

(a) Zero
(b) .0107
(c) .9893
(d) .5000
(e) .4893

Answer: (b)
Difficulty level: Easy to Medium

7.7.5: If X and Y are random variables with a <u>positive</u> covariance, which of the following *must* be true?

(a) Variance(X - Y) < [Variance(X) + Variance(Y)].
(b) Variance(X - Y) > [Variance(X) + Variance(Y)].
(c) Variance(X - Y) = [Variance(X) + Variance(Y)].
(d) Variance(X - Y) = Variance(X + Y).
(e) More than one of these must be true.

Answer: (a)
Difficulty level: Medium

7.7.6: If the populations from which the samples are drawn *are not* normal in shape, then the sampling distribution of the difference in sample means will:

(a) be normally distributed by the Central limit theorem.
(b) be normally distributed, but with an unknown mean and variance.
(c) be normally distributed, regardless of how large or small the samples sizes are.
(d) be normally distributed if and only if both sample sizes are sufficiently large.
(e) never be normally distributed, regardless of how large or small the samples sizes are.

Answer: (d)
Difficulty level: Easy

7.8.1: Airlines often overbook flights in order to minimize the effects of people who have reservations but fail to show up for flights. The fraction of no-shows for two airlines are 10% for Airline 1 and 12% for Airline 2. Each airline randomly selects flights with 120 reservations. What is the probability that the fraction of no-shows for the two airlines will differ by an amount that is *greater than* ± 1%?

(a)   .8026
(b)   .6310
(c)   .3690
(d)   .1717
(e)   .8283

Answer: (e)
Difficulty level: Medium to Challenging

7.8.2: Airlines often overbook flights in order to minimize the effects of people who have reservations but fail to show up for flights. The fraction of no-shows for two airlines are 10% for Airline 1 and 12% for Airline 2. Each airline randomly selects flights with 120 reservations. What is the probability that the fraction of no-shows for the two airlines, will differ by an amount that is EITHER larger than ± 4% or smaller than ± 1%?

(a)   .4518
(b)   .5482
(c)   .7950
(d)   .0404
(e)   .5148

Answer: (b)
Difficulty level: Challenging

7.8.3: Two different salespeople working at Mutual of New York (MONY) have sales success rates of $p_1 = .12$ and $p_2 = .08$ when they "cold call" customers. If the two salespeople both called 250 randomly selected customers during the past month, what is the probability that the sales success rate of salesperson 1 is at least 5% higher than the sales success rate of salesperson 2?

(a)   .1443
(b)   .8557
(c)   .6443
(d)   .3557
(e)   .0307

Answer: (d)
Difficulty level: Easy to Medium

7.8.4: Two different salespeople working at Mutual of New York (MONY) have sales success rates of $p_1 = .12$ and $p_2 = .08$ when they "cold call" customers. If the two salespeople both called 250 randomly selected customers during the past month, what is the probability that salesperson 1's sales success rate is NEITHER at least as high NOR at least 5% lower than the sales success rate of salesperson 2?

(a)   .0677
(b)   .9323
(c)   .9315
(d)   .4323
(e)   .5687

Answer: (a)
Difficulty level: Medium to Challenging

7.8.5: Southern Industries, Inc., has two sales outlets. At both outlets, 40% of the customers charge their purchases. During a recent audit of 100 sales slips from each outlet, the company accountant found 44 and 36 charge customers from the first and second outlets, respectively. What is the probability that a result would be achieved whereby the first outlet's proportion of charge customers exceeded that of the second by an amount that is *at least* this much?

(a)   .1251
(b)   .3749
(c)   .8749
(d)   .0693
(e)   .5000

Answer: (a)
Difficulty level: Medium

Estimation

8.1.1: Sample statistics, such as the sample mean and the sample proportion, that are used to estimate the value of unknown population parameters are called:

(a)   minimum-variance unbiased or best estimators
(b)   consistent estimators
(c)   unbiased estimators
(d)   estimators
(e)   all of these

Answer: (d)
Difficulty level: Easy

8.1.2: The sample mean is often used as an estimator of the population mean because of desirable properties, including:

(a)   unbiasedness and efficiency only
(b)   consistency and minimum sampling error only
(c)   unbiasedness and consistency only
(d)   unbiasedness, consistency, and efficiency
(e)   unbiasedness, consistency, and minimum sampling error

Answer: (d)
Difficulty level: Easy

8.1.3: In statistics, a *best* estimator is one that:

(a)   has the smallest sampling error among estimators
(b)   has the smallest variance among estimators
(c)   is an unbiased estimator
(d)   has the largest variance among unbiased estimators
(e)   has the smallest variance among unbiased estimators

Answer: (e)
Difficulty level: Easy

8.1.4: The sample proportion has which of the following properties when used to estimate the population proportion?

(a)   unbiasedness
(b)   consistency
(c)   unbiasedness and minimum sampling error
(d)   unbiasedness, consistency, and efficiency
(e)   unbiasedness, consistency, and minimum sampling error

Answer: (d)
Difficulty level: Easy to Medium

8.1.5: If the expected value of a sample statistic is equal to the population parameter for which the statistic is an estimator, this means that the statistic (or estimator):

(a)   is unbiased
(b)   is consistent
(c)   is efficient
(d)   minimizes the sampling error
(e)   is both unbiased and consistent

Answer: (a)
Difficulty level: Easy

8.1.6: If the variance of a sample statistic is the smallest among all unbiased estimators of a given population parameter, this means that the statistic (or estimator):

(a)   is unbiased.
(b)   is efficient.
(c)   minimizes the sampling error.
(d)   is a maximum-variance or best estimator.
(e)   is a minimum-variance unbiased or best estimator.

Answer: (e)
Difficulty level: Easy

8.1.7: The sample standard deviation is:

(a)   a biased but consistent estimator of σ
(b)   a biased and inconsistent estimator of σ
(c)   an unbiased and consistent estimator of σ
(d)   a minimum-variance unbiased estimator of σ
(e)   an unbiased and inconsistent estimator of σ

Answer: (a)
Difficulty level: Medium

8.1.8: A *point estimate* is defined as:

(a)   the midpoint of values observed for the statistic
(b)   a lower and upper bound for the population parameter
(c)   the number that represents the best estimate of the population parameter
(d)   a lower and upper bound for a sample statistic
(e)   a minimum-variance unbiased estimate of a parameter

Answer: (c)
Difficulty level: Easy to Medium

8.1.9: Since its' sampling distribution collapses around the population mean and its' variance approaches zero as n increases toward infinity, the sample mean is said to be a(n) _____ estimator of the population mean.

(a) unbiased
(b) efficient
(c) consistent
(d) asymptotically unbiased
(e) minimum-variance unbiased or best

Answer: (c)
Difficulty level: Easy to Medium

8.1.10: If the population is known to be *skewed*, then $\bar{X}$ is a more desirable estimator of the population mean μ than is the sample median because the:

(a) sample mean is a consistent estimator of μ
(b) sample median does not minimize the mean square error
(c) sample median is a biased estimator of μ
(d) sample median does not minimize the sampling error
(e) none of these

Answer: (c)
Difficulty level: Medium

8.1.11: When used as an estimator of σ, the sample standard deviation, s, *does not* have the property of:

(a) unbiasedness
(b) efficiency
(c) consistency
(d) asymptotic unbiasedness
(e) s has all of these properties when used to estimate σ

Answer: (a)
Difficulty level: Medium

8.1.12: For the following results from a sample taken from a population, the point estimates for μ and σ^2 are:

$$n = 9 \qquad \Sigma \ X_i = 36 \qquad \Sigma \ (X_i - \bar{X})^2 = 288$$

(a) 4.0 and 32.0, respectively
(b) 4.0 and 36.0, respectively
(c) 4.0 and 6.0, respectively
(d) 4.0 and 4.0, respectively
(e) 4.0 and 2.0, respectively

Answer: (b)
Difficulty level: Easy

8.1.13: For the following results from a sample taken from a population, the point estimates for the expected value and standard deviation of the sample mean are:

$$n = 9 \qquad \Sigma \ X_i = 36 \qquad \Sigma \ (X_i - \bar{X})^2 = 288$$

(a) 4.0 and 1.89, respectively
(b) 4.0 and 36.0, respectively
(c) 4.0 and 6.0, respectively
(d) 4.0 and 4.0, respectively
(e) 4.0 and 2.0, respectively

Answer: (e)
Difficulty level: Medium

8.1.14: For the following results from a sample taken from a population, the point estimates for μ and σ are:

$$n = 25 \qquad \Sigma \ X_i = 500 \qquad \Sigma \ (X_i)^2 = 12,400$$

(a) 20.0 and 100, respectively
(b) 20.0 and 22.73, respectively
(c) 20.0 and 10.0, respectively
(d) 20.0 and 4.0, respectively
(e) 20.0 and 2.0, respectively

Answer: (c)
Difficulty level: Easy

8.1.15: For the following results from a sample taken from a population, the point estimates for the expected value and standard deviation of the sample mean are:

$$n = 25 \qquad \Sigma \; X_i = 500 \qquad \Sigma \; (X_i)^2 = 12,400$$

(a)   20.0 and 22.73, respectively
(b)   20.0 and 4.55, respectively
(c)   20.0 and 10.0, respectively
(d)   20.0 and 4.0, respectively
(e)   20.0 and 2.0, respectively

Answer: (e)
Difficulty level: Medium

8.1.16: A real estate broker for Geselman and Associates randomly selected four properties from a subdivision, and an appraiser gave an opinion on the value of each property as $17, $21, $21, and $21 thousand. The point estimates for the mean and variance of the property values in the subdivision, μ and σ^2, are:

(a) $20,000 and $3,000, respectively
(b) $20,000 and $3,000,000, respectively
(c) $20,000 and 3,000,000 (squared dollars), respectively
(d) $20,000 and 3,000 (squared dollars), respectively
(e) $20,000 and 3 (squared dollars), respectively

Answer: (c)
Difficulty level: Medium

8.2.1: The reason for constructing a confidence interval instead of relying on a point estimate of a parameter is:

(a)   interval estimates are more precise than point
       estimates
(b)   interval estimates are both more accurate and more
       precise than point estimates
(c)   a confidence interval can be used to test hypotheses
       about the parameter
(d)   a confidence interval provides an idea of how close the
       point estimate is likely to be to the parameter
(e)   a confidence interval represents a lower and upper
       bound for the population parameter with a probability
       that the estimate falls within that range being the
       confidence level of $(1 - \alpha)$

Answer: (d)
Difficulty level: Easy to Medium

8.2.2: A reason why confidence interval estimates for population parameters are preferred to point estimates is:

(a) interval estimates are more precise than point estimates
(b) interval estimates are less accurate than point estimates
(c) interval estimates are both more accurate and more precise than point estimates
(d) interval estimates take into account the fact that the statistic (or estimator) being used to estimate the population parameter is a random variable
(e) more than one of these

Answer: (d)
Difficulty level: Easy to Medium

8.2.3: A *confidence interval* is defined as:

(a) a lower and upper bound for the population parameter
(b) a minimum-variance unbiased estimate of the population parameter
(c) the number that represents the best estimate of the population parameter
(d) a lower and upper bound for the population parameter with an associated confidence level of $(1 - \alpha)$
(e) a lower and upper bound for the population parameter with a probability that the estimate falls within that range being the confidence level of $(1 - \alpha)$

Answer: (d)
Difficulty level: Easy to Medium

8.2.4: A 90% *confidence interval* for p means that:

(a) there is a 90% probability that the sample mean $\overline{X}$ will lie in between the lower and upper bounds of the confidence interval   ^
(b) there is a 90% probability that the sample proportion p will lie in between the lower and upper bounds of the confidence interval
(c) approximately 90% of similarly constructed intervals will contain the true proportion p
(d) approximately 90% of similarly constructed intervals will contain the true mean μ
(e) you are 90% confident that the true proportion p is contained within the confidence interval

Answer: (c)
Difficulty level: Easy to Medium

8.2.5: When constructing a confidence interval for μ, the finite population correction factor *must* be used when:

(a) sampling is done without replacement
(b) the sample size is at least 5% of the population
(c) sampling is done without replacement *or* the sample size is at least 5% of the population
(d) sampling is done without replacement *and* the sample size is at least 5% of the population
(e) the finite population correction factor is *never* used when constructing confidence intervals for μ

Answer: (d)
Difficulty level: Easy to Medium

8.2.6: The circumstance of estimating the population mean μ of a normal population when the standard deviation σ is known is:

(a) typical because one almost always knows what σ when μ is unknown
(b) typical because one rarely knows what σ is when μ is unknown
(c) not typical because one almost always knows what σ is when μ is unknown
(d) not typical because one rarely knows what σ is when μ is unknown
(e) not typical because one usually knows what μ is but does not know what σ is

Answer: (d)
Difficulty level: Easy to Medium

8.2.7: The most common situation when constructing a confidence interval for the population mean is:

(a) a non-normal population with a small sample
(b) a non-normal population with an unknown population standard deviation
(c) a normal population with a known population standard deviation
(d) a normal population with a small sample
(e) a normal population with an unknown population standard deviation

Answer: (b)
Difficulty level: Medium

8.2.8: A sample of size n = 9 is taken from a *normally distributed* population with a known standard deviation σ of 45. If the sample mean equals 150, what is the upper limit of a 95% confidence interval for the population mean?

(a)   29.4
(b)   120.6
(c)   179.4
(d)   125.25
(e)   174.75

Answer: (c)
Difficulty level: Easy to Medium

8.2.9: A sample of size n = 16 is taken from a *normally distributed* population with a known standard deviation σ of 40. If the sample mean equals 200, the lower limit of a 95% confidence interval for the population mean is:

(a)   19.6
(b)   180.4
(c)   219.6
(d)   183.5
(e)   216.5

Answer: (b)
Difficulty level: Easy to Medium

8.2.10: A sample of size n = 16 is taken from a *normally distributed* population with a known standard deviation of 80. If the sample mean is 250, the lower and upper limits of a 98% confidence interval for the population mean are:

(a)   203.40 to 296.60
(b)   238.35 to 261.65
(c)   198.40 to 301.60
(d)   237.10 to 262.90
(e)   none of these

Answer: (a)
Difficulty level: Easy to Medium

8.2.11: A pharmaceutical company has a sleep-enhancing drug on the market. The company needs to estimate the drug's effectiveness in terms of the mean amount of increase in sleep that patients might expect. If the drug is an effective sleep-enhancer, then the profits from the sales of the drug are likely to increase the company's stock price. A sample of 10 patients was done, with the results for the amount of increase in sleep (in hours) for each patient summarized below. Assume we *know from past experience* that the increase in sleep is normally distributed.

Patient	Increase	Patient	Increase
1	1.2	6	1.0
2	2.4	7	1.8
3	1.3	8	0.8
4	1.3	9	4.6
5	0.0	10	1.4

a. Given a mean additional hours of sleep gained by using the drug of 1.58 hours, what is the sample standard deviation s?
b. Given a mean additional hours of sleep gained by using the drug of 1.58 hours, construct a 95% confidence interval for μ using a known population standard deviation of σ = 1.29 hours.
c. Given a mean additional hours of sleep gained by using the drug of 1.58 hours, construct a 90% confidence interval for the amount of increase in sleep μ that uses the sample standard deviation from part a.
d. Given a mean additional hours of sleep gained by using the drug of 1.58 hours, construct a 98% confidence interval for the amount of increase in sleep μ that uses the sample standard deviation from part a.
e. Compare the width and risk of reporting an incorrect interval for the confidence intervals in parts c and d.

Answer: a.   s = (1/9)[13.616] = 1.51 hours
           b.   1.58 ± (1.96)(1.29/square root of 10)
                [0.78 to 2.38 hours of additional sleep]
           c.   1.58 ± (1.833)(1.51/square root of 10)
                [0.70 to 2.46 hours of additional sleep]
           d.   1.58 ± (2.821)(1.51/square root of 10)
                [0.23 to 2.93 hours of additional sleep]
           e.   The width is narrower and the risk of reporting an incorrect interval is higher for the first confidence interval
Difficulty level: Medium

8.2.12: The $100(1 - \alpha)$% **confidence interval for the population mean** μ when the population is normally distributed and the standard deviation σ is known is an interval with an *upper* confidence limit of:

(a) $L = \overline{X} + Z_{\alpha}\sigma_X/$[square root of n]

(b) $R = \overline{X} - Z_{\alpha/2}\sigma_X/$[square root of n]

(c) $L = \overline{X} + t_{\alpha/2,n-1}s_X/$[square root of n]

(d) $R = \overline{X} - t_{\alpha/2,n-1}s_X/$[square root of n]

(e) $L = \overline{X} + Z_{\alpha/2}\sigma_X/$[square root of n]

Answer: (e)
Difficulty level: Easy

8.2.13: The $100(1 - \alpha)$% **confidence interval for the population mean** μ when the population is normally distributed and the standard deviation σ is unknown is an interval with a *lower* confidence limit of:

(a) $L = \overline{X} + Z_{\alpha}\sigma_X/$[square root of n]

(b) $R = \overline{X} - Z_{\alpha/2}\sigma_X/$[square root of n]

(c) $L = \overline{X} + t_{\alpha/2,n-1}s_X/$[square root of n]

(d) $R = \overline{X} - t_{\alpha/2,n-1}s_X/$[square root of n]

(e) $L = \overline{X} + Z_{\alpha/2}\sigma_X/$[square root of n]

Answer: (d)
Difficulty level: Easy

8.2.14: The confidence limits for a 95% confidence interval for the true mean when X is normal, given a sample mean of 25, a known standard deviation of $\sigma = 5$, and n = 25 are:

(a)   23.355 and 26.645
(b)   24.608 and 25.392
(c)   24.484 and 25.516
(d)   23.04 and 26.96
(e)   22.42 and 27.58

Answer: (d)
Difficulty level: Easy

8.3.1: Find the value of t for the following situations:

a.   The probability of a smaller value is .025 when the
      sample size is 25.
b.   The probability of a larger value is .01 when the
      sample size is 10.
c.   The probability of a larger value is .90 when the
      sample size is 41.
d.   The probability that a value lies in between plus or
      minus this t value is .95 when the sample size is 20
e.   The probability that a value lies in between plus or
      minus this t value is .99 when the sample size is 9

Answer: a.   t = -2.064
           b.   t = 2.821
           c.   t = -1.310
           d.   t = 2.093
           e.   t = 3.355
Difficulty level: Easy to Medium

8.3.2: A sample of size n = 16 is taken from a *normally
distributed* population with an unknown standard deviation.
If the sample mean and standard deviation were found to be
400 and 16, the upper limit of a 95% confidence interval is:

(a)   8.52
(b)   391.48
(c)   408.52
(d)   392.99
(e)   407.01

Answer: (c)
Difficulty level: Easy to Medium

8.3.3: A sample of size n = 16 is taken from a *normally
distributed* population with an unknown standard deviation.
If the sample mean and sample standard deviation were found
to be 400 and 16, respectively, the appropriate critical Z
(or t) values that correspond to the lower and upper limits
of a 99% confidence interval for the population mean are:

(a) Z_α =   + 2.33
(b) $Z_{\alpha/2}$ =   ± 2.58
(c) $t_{\alpha/2}$ =   ± 2.58
(d) $t_{\alpha, n-1}$ =   + 2.602
(e) $t_{\alpha/2, n-1}$ =   ± 2.947

Answer: (e)
Difficulty level: Easy

8.3.4: Average grade point averages at a prestigious college in the eastern United States are known from past experience to be normally distributed. A sample of n = 25 students is taken with a sample mean of 2.30 and a standard deviation of 0.75. The lower and upper limits of a 95% confidence interval for the population mean are:

(a) 1.99 and 2.61
(b) 1.55 and 3.05
(c) 0.80 and 3.80
(d) 2.238 and 2.362
(e) 2.043 and 2.857

Answer: (a)
Difficulty level: Easy to Medium

8.3.5: When constructing confidence intervals for μ, it is optional to use the t score when:

(a) the population is normal and σ is known
(b) the population is skewed and σ is known
(c) the population is normal, the population standard deviation is unknown, and the sample size is small
(d) the population is normal, the population standard deviation is unknown, and the sample size is large
(e) the population is normal, the population standard deviation is known, and the sample size is large

Answer: (d)
Difficulty level: Medium

8.3.6: Which of the following assumptions *must* be satisfied in order for the t distribution to be well defined?

(a) the population is normally distributed
(b) the sample size is sufficiently large
(c) the population standard deviation is known
(d) the sample standard deviation is an unbiased estimator of the population standard deviation
(e) none of these

Answer: (a)
Difficulty level: Easy to Medium

8.3.7: The reason that the t score is used in place of the Z score when the population standard deviation is unknown is:

(a)   the t score is always lower than the Z score
(b)   the t score has fewer degrees of freedom
(c)   the t score gives you a more precise or narrower confidence interval
(d)   the t score adjusts for the bias in the sample standard deviation
(e)   the t score is never used in place of the Z score

Answer: (d)
Difficulty level: Medium

8.4.1: A random sample of 20 homeowners in Akron, Ohio, showed that the mean monthly mortgage payment being made by the people in the sample was $800 with a sample standard deviation of $60. Assuming a normal population, the lower and upper confidence limits of a 98% confidence interval for the mean monthly mortgage payment are:

(a) $680 and $920
(b) $765.9 and $834.1
(c) $792.4 and $807.6
(d) $768.7 and $831.3
(e) $770.9 and $829.1

Answer: (b)
Difficulty level: Easy to Medium

8.4.2: A random sample of 28 families in New York City revealed that the children in these families averaged viewing 130 minutes of television each day with a sample standard deviation of 50 minutes. Assuming a *normally distributed* population, the lower and upper confidence limits of a 98% confidence interval for the mean daily viewing time of children in New York City are:

(a)   106.63 and 153.37 minutes
(b)   107.98 and 152.02 minutes
(c)   119.5 and 150.50 minutes
(d)   120.61 and 149.39 minutes
(e)   125.58 and 134.42 minutes

Answer: (a)
Difficulty level: Easy to Medium

8.4.3: A retailer surveys 100 of his customers during a sale and determines that the average purchase was $8.73 with a sample standard deviation of $1.00. The lower and upper confidence limits for a 95% confidence interval for μ are:

(a) $6.77 and $10.69
(b) $8.53 and $8.93
(c) $6.75 and $10.71
(d) $7.73 and $9.73
(e) $6.73 and $10.73

Answer: (b)
Difficulty level: Easy to Medium

8.4.4: A quality control engineer took a sample of n = 81 items coming off a production line and found that on the average they weighed 15.5 pounds with a sample standard deviation of s = 2.7 pounds. The *upper* confidence limit of a 95% confidence interval for the population mean of item weights coming from this production line is:

(a) 16.088 pounds
(b) 16.094 pounds
(c) 15.565 pounds
(d) 15.995 pounds
(e) 15.800 pounds

Answer: (b)
Difficulty level: Medium

8.4.5: The *lower* confidence limit of a 99% confidence interval for μ given a normal population, a sample mean of 8, a sample standard deviation of 4, and n = 4 is:

(a) -3.68
(b) 19.68
(c) 3.34
(d) 2.84
(e) 0

Answer: (a)
Difficulty level: Easy to Medium

8.4.6: The lower and upper limits of a 90% confidence interval for μ given a normal population, a sample mean of 8, a sample standard deviation of 4, and n = 25 are:

(a)   6.944 and 9.056
(b)   6.631 and 9.369
(c)   6.349 and 9.651
(d)   -1.968 and 17.968
(e)   -3.188 and 19.188

Answer: (b)
Difficulty level: Easy to Medium

8.5.1: A random sample of 100 individuals was taken to determine the true percentage of people who smoke in a region of the eastern United States. The 100 people were asked to indicate whether or not they were smokers. Forty-six of them said "yes." The lower and upper confidence limits of a 95% confidence interval for the true proportion of <u>nonsmokers</u> are:

(a)   .46 and .54
(b)   .36 and .56
(c)   .95 and 1.0
(d)   .44 and .64
(e)   .00 and .95

Answer: (d)
Difficulty level: Medium

8.5.2: Find the *upper* confidence limit of a 95% confidence interval for the proportion of employees in the health care industry who are concerned about acquiring an immune deficiency virus on the job if a random sample of 250 workers reveals that 150 expressed concern about acquiring such a virus on the job.

(a)   .631
(b)   .651
(c)   .661
(d)   .569
(e)   .539

Answer: (c)
Difficulty level: Easy to Medium

8.5.3: Auto Motor Company has received a shipment of several thousand parts. A random sample of 81 parts is selected, 8 of which are found to be defective. The *upper* confidence limit of a 90% confidence interval for the true proportion of defects in the entire shipment of parts is:

(a) .0988
(b) .0563
(c) .1413
(d) .1533
(e) .0443

Answer: (d)
Difficulty level: Easy to Medium

8.5.4: If a random sample of 1000 found the number of businesses in favor of raising taxes to reduce the debt is 50, the *upper* confidence limit of a 95% confidence interval for the true proportion of small businesses in favor of higher tax rates to reduce the national debt is:

(a) .057
(b) .061
(c) .072
(d) .064
(e) .028

Answer: (d)
Difficulty level: Easy to Medium

8.5.5: If a random sample of 100 employees found 64 to be enrolled in the health insurance organization's medical plan, the *lower* confidence limit of a 95% confidence interval for the true proportion of employees enrolled in such a health insurance plan is:

(a) .688
(b) .719
(c) .734
(d) .546
(e) .512

Answer: (d)
Difficulty level: Easy to Medium

8.6.1: The credit manager of a department store would like to know what proportion of the charge customers take advantage of the store's deferred-payment plan each year. She would like to estimate this proportion within ± .10 at a 90% confidence level, but she has no good idea about what this proportion might be. How many customers should she sample?

(a)   271
(b)   41
(c)   17
(d)   68
(e)   Cannot be determined

Answer: (d)
Difficulty level: Easy to Medium

8.6.2: If a normal population of overpayments by National Insurance to a health service provider is known to have a population standard deviation equal to $10, how large a sample would be need to take in order to be 95% confident that the sample mean overpayment will not differ from the population mean overpayment by more than ± $1.00?

(a)   97
(b)   1,537
(c)   385
(d)   273
(e)   Cannot be determined

Answer: (c)
Difficulty level: Easy to Medium

8.6.3: A quality control engineer with Systems Planning Corporation would like to estimate the proportion of defects being produced on a production to within ± .04 with 95% confidence. If a preliminary sample of n = 25 items reveals 8 defectives, how many *additional* items need to be sampled?

(a)   523
(b)   131
(c)   498
(d)   346
(e)   Cannot be determined

Answer: (c)
Difficulty level: Medium

8.6.4: Which of the following *is not* necessary information for determining the minimum sample size for confidence intervals?

(a)   the confidence level
(b)   the desired maximum allowable margin for sampling error
(c)   the standard deviation of the population
(d)   the mean of the population
(e)   the value for α

Answer: (d)
Difficulty level: Easy

8.6.5: Using a 99% confidence level and a planning value of 0.018, what is the minimum sample size needed to estimate the true proportion of residents in the eastern United States who carry the AIDS virus within ± 2%?

(a)   240
(b)   1,177
(c)   74
(d)   295
(e)   Cannot be determined

Answer: (d)
Difficulty level: Easy

8.6.6: A preliminary sample of 10 ball bearings made by a Low-Shear Industries process yielded a standard deviation of .30 millimeters. How many *additional* ball bearings would need to be sampled in order to estimate the mean diameter of all ball bearings made by this process to within ± .01 millimeters with a confidence level of 95%?

(a)   4,605
(b)   1,152
(c)   4,595
(d)   3,014
(e)   Cannot be determined

Answer: (c)
Difficulty level: Easy to Medium

8.6.7: A minimum sample size needed to estimate a population mean within ± 5 at a 95% confidence level with a planning value for the standard deviation of 40 is:

(a)   62
(b)   44
(c)   1,537
(d)   175
(e)   246

Answer: (e)
Difficulty level: Easy

8.6.8: A minimum sample size needed to estimate a population mean within ± 5 at α = .05, given σ = 20, is:

(a)   62
(b)   44
(c)   8
(d)   1,537
(e)   246

Answer: (a)
Difficulty level: Easy

8.6.9: An accountant wants to estimate the average dollar amount of taxes paid by single people aged 20 to 25. Since the accountant handles a large number of people in that age group, she decides to take a sample. If she believes that the true standard deviation is equal to $1500, then how large a sample should she take if she wants to be 95% confident that her sample mean will differ from the true mean by no more than ± $200?

(a)   55
(b)   153
(c)   217
(d)   865
(e)   Cannot be determined

Answer: (c)
Difficulty level: Easy

8.6.10: A research firm wants to estimate the true proportion of families having annual incomes that exceed $100,000 in a particular city. Using a planning value of 20%, how large a sample should the research firm use to be 98% confident that the sample proportion differs from the true proportion by no more than ± 3%.

(a)   61
(b)   242
(c)   966
(d)   1,509
(e)   3,861

Answer: (c)
Difficulty level: Easy

8.7.1: The standard error of the mean plays an important role in inferential statistics because:

(a)   it detects mistakes in the arithmetic mean
(b)   it detects and corrects for sampling errors
(c)   it tells how large the mean should be
(d)   it tells how much sample means can be expected to vary from sample to sample
(e)   it tells us how large the sample size should be

Answer: (d)
Difficulty level: Easy to Medium

8.7.2: From the formula for the standard error of the mean, we know that the standard error of the mean *decreases* when:

(a)   we decrease the sample size n
(b)   we increase the standard deviation σ
(c)   we combine an increase in both σ and n
(d)   we combine an increase in σ with a decrease in n
(e)   we combine a decrease in σ with an increase in n

Answer: (e)
Difficulty level: Easy to Medium

8.8.1: When dealing with samples from two populations, the best unbiased estimator for the difference in sample variances when the common standard deviation $\sigma = \sigma_1 = \sigma_2$ is *unknown* is:

(a) the pooled degrees of freedom
(b) the pooled sample variance
(c) the sum of the sample variances
(d) the sample standard deviation
(e) there is no such thing

Answer: (b)
Difficulty level: Easy

8.8.2: To obtain an unbiased estimate for the common population variance from two samples:

(a) we calculate the pooled estimate of variance.
(b) we add the sum of squares from each sample and then divide by the pooled degrees of freedom.
(c) we get the weighted mean of the individual sample estimates, where the weights are the degrees of freedom.
(d) more than one of these is correct.
(e) an educated guess is the best approach.

Answer: (d)
Difficulty level: Easy to Medium

8.8.3: In an experiment designed to compare the means of normally distributed service lives of two types of tires at Eagle, Inc., the difference between the sample means was 2,250 miles, and the pooled estimate of the sample variance, based on samples of 10 tires of each type, was equal to 625,000. The lower and upper confidence limits of a 95% confidence interval for the true difference between the mean lives of the two types of tires are:

(a) 1,896.45 and 2,603.55 miles
(b) 1,636.94 and 2,863.06 miles
(c) 1,507.19 and 2,992.81 miles
(d) 1,668.41 and 2,831.59 miles
(e) 1,557.04 and 2,942.96 miles

Answer: (c)
Difficulty level: Medium

8.8.4: Independent random samples of size 7 from normal populations with equal standard deviations of the amount of insider trading by employees during a merger of (1) investment banking firms and (2) brokerage houses gave a sample difference in means of +4.0, with sample variances of 4.0 and 2.0. The *upper* confidence limit of a 95% confidence interval for the true difference in means $\mu_1 - \mu_2$ is:

(a)   5.650
(b)   7.312
(c)   7.683
(d)   2.317
(e)   6.018

Answer: (e)
Difficulty level: Medium

8.8.5: Independent random samples of size 7 from normal populations with equal standard deviations of the breaking load for two fabricated wooden beams. The sample mean and standard deviation for beam 1 were 2,000 and 900 pounds, and were 1,500 and 600 pounds for beam 2. The *upper* confidence limit of a 99% confidence interval for the true difference in mean breaking loads for the wooden beams $\mu_1 - \mu_2$ is:

(a)   1,572.77 pounds
(b)   1,552.74 pounds
(c)   1,717.09 pounds
(d)   1,748.98 pounds
(e)   Cannot be determined

Answer: (d)
Difficulty level: Medium

8.9.1: In a sample of 100 from one binomial population there were 28 successes. A sample of 200 from a second binomial population had 92 successes.

a.   Find a 90% confidence interval for $p_1 - p_2$.
b.   Find a 98% confidence interval for $p_1 - p_2$.
c.   Compare the width (W) and risk of reporting an incorrect interval (α) of the two confidence intervals.

Answer: a.   -.274 to -.086
           b.   -.313 to -.047
           c.   The first confidence interval for $p_1 - p_2$
                 (from part a) is narrower and has a higher
                 risk than the second one (from part b)
Difficulty level: Easy to Medium

Tests of Hypotheses

9.1.1: Determine the appropriate null and alternative hypotheses for each of the following and state whether the hypothesis test should be one-sided or two-sided.

a. A quality control manager is concerned that the process for a particular bolt cutting machine is not "in control." The machine is "in control" if bolts have a length that lies within the range .50 ± .02 inches.

b. The Food and Drug Administration (FDA) has a known standard for pizza sauce that specifies that at most .01 milligrams of a food preservative may be contained in a given can of pizza sauce. For cans that contain more than this amount, the FDA may recall some of the company's product from the market.

c. An accounting major is considering enrolling in a training program whose director maintains from past experience at least 90% of all accountants who take the training program pass the Certified Public Accounting (CPA) exam. She would like to know whether the advertising claim is now false.

d. A quality control engineer is interested in knowing whether the proportion of special photographic lenses that fail to pass inspection is currently the same at two manufacturing plants that produce the lenses. Historically, the proportion of lenses that fail to pass inspection has been the same at the two plants.

e. A company training director would like to know whether there is a significant difference in the productivity of workers trained in two different training programs to determine whether more workers should be trained under the new training program.

Answer: a. H_0: $\mu = .50"$; H_a: $\mu \neq .50"$; two-sided

b. H_0: $\mu \leq .01$ mg; H_a: $\mu > .01$ mg; one-sided

c. H_0: $p \geq .90$; H_a: $p < .90$; one-sided

d. H_0: $p_1 = p_2$; H_a: $p_1 \neq 0$; two-sided

e. H_0: $\mu_1 = \mu_2$; H_a: $\mu_1 \neq \mu_2$; two-sided

Difficulty level: Easy to Medium

9.1.2: Which of the following statements about hypothesis tests are *false*?

(a)   Hypothesis tests can be either one-sided or two-sided
(b)   If done correctly, hypothesis tests always lead to the right conclusion
(c)   Hypothesis tests can be performed for population means and population proportions
(d)   Hypothesis tests can be performed to test for the equality of two unknown population variances
(e)   All of these statements are false

Answer: (b)
Difficulty level: Easy

9.1.3: The null hypothesis is often established in such a way that:

(a)   the status quo is always maintained
(b)   it states nothing is different from what it is supposed to be or has been in the past
(c)   the alternative hypothesis is always true
(d)   the alternative hypothesis is always false
(e)   Type I and Type II errors are both minimized

Answer: (b)
Difficulty level: Easy

9.1.4: Rejection of the null hypothesis implies:

(a)   selection of another null hypothesis
(b)   acceptance of the alternative hypothesis
(c)   presence of a sampling error
(d)   weak evidence in support of the null hypothesis
(e)   none of these

Answer: (b)
Difficulty level: Easy

9.1.5: Failure to reject the null hypothesis implies that:

(a)   the null hypothesis must be true
(b)   the alternative hypothesis must be false
(c)   a Type I error might have occurred
(d)   a Type II error might have occurred
(e)   there is strong evidence supporting the null hypothesis

Answer: (d)
Difficulty level: Easy to Medium

9.2.1: A *rejection rule*:

(a) indicates the values of the sample or test statistic that will lead us to reject a null hypothesis
(b) indicates the values of the sample or test statistic that will lead us to not reject a null hypothesis
(c) can be either one-sided or two-sided, depending on whether the hypothesis test is one-sided or two-sided
(d) can be either one-sided or two-sided, depending on whether the sample or test statistic is positive or negative
(e) more than one of these statements about rejection rules is correct

Answer: (e)
Difficulty level: Easy

9.2.2: In a one-tailed test situation, α represents:

(a) the probability of a Type I error occurring when the null hypothesis is false
(b) the maximum probability of a Type II error
(c) the correct decision
(d) the probability of a Type I error occurring when the parameter is equal to the value indicated by the null hypothesis
(e) the minimum probability of a Type II error

Answer: (d)
Difficulty level: Easy

9.2.3: In tests of significance, if the difference between what we expect and what we get is so large that it cannot reasonably be attributed to chance, we say that our results:

(a) are known to be reliable
(b) are known to be bad
(c) are proven
(d) are statistically significant
(e) are not statistically significant

Answer: (d)
Difficulty level: Easy

9.3.1: If one cannot reject H_0: $\mu = 6$ at $\alpha = .01$, then one:

(a)   must reject H_0 for any value of α greater than .01
(b)   cannot reject H_0 for any value of α greater than .01
(c)   might reject H_0 if the value of α is increased
(d)   might reject H_0 if the sample size n is increased
(e)   has proven that H_0 is true beyond a reasonable doubt

Answer: (c)
Difficulty level: Medium

9.3.2: Larger values of α are tolerated when:

(a)   we are not really interested in the outcome
(b)   a Type I error is not as serious as a Type II error
(c)   we desire a low probability of erroneously rejecting H_0
(d)   the cost of committing a Type II error is very low
(e)   larger values of α are never tolerated

Answer: (b)
Difficulty level: Easy to Medium

9.3.3: Knowledge of Type I and Type II errors is useful:

(a)   because we can assess the strength of a test or a
      rejection region by knowing about the probabilities of
      these two kinds of errors
(b)   only when we are unsure which error is more costly
(c)   when we do not have reliable and representative samples
(d)   because there is always the possibility of committing
      either a Type I or a Type II error, regardless of
      whether the null hypothesis is true
(e)   knowledge of Type I and Type II errors is never useful

Answer: (a)
Difficulty level: Easy

9.3.4: In a United States courtroom setting, convicting an
innocent person is an example of:

(a)   justice
(b)   a biased jury
(c)   an honest mistake
(d)   a type I error
(e)   a type II error

Answer: (d)
Difficulty level: Medium

9.3.5: In a United States courtroom setting, rules such as the Fifth Amendment that allow defendants to avoid self-incrimination, thus making it more difficult to convict an accused defendant, will in general:

(a) have no affect on either type I or type II errors
(b) have no affect on type I errors, but will increase the probability of committing a type II error
(c) increase the probability of committing both type I and type II errors
(d) increase the probability of committing a type I error and decrease the probability of committing a type II error
(e) decrease the probability of committing a type I error and increase the probability of committing a type II error

Answer: (e)
Difficulty level: Medium

9.3.6: In screening for possible cancer cases, the null hypothesis (H_0) is "This patient does not have cancer." Which of the following is the most dangerous from the patients' perspective?

(a) a Type I error
(b) H_0 being true when it is accepted.
(c) H_0 being true when it is rejected.
(d) H_0 being false when it is rejected.
(e) H_0 being false when it is accepted.

Answer: (e)
Difficulty level: Easy to Medium

9.3.7: It is commonly believed that students with low LSAT scores will not do well in law school because they are not "bright enough." An admittance committee may thus decide to reject a student (H_1) on the basis of the LSAT score. If the committee does this, it has _____ if the student is in fact "bright enough" to attend law school.

(a) committed a type I error
(b) committed a type II error
(c) made the right decision
(d) committed a sampling error
(e) none of these

Answer: (a)
Difficulty level: Easy

9.3.8: The level of significance or size of the test is
denoted by ____. The power of the test is denoted by ____.

(a) α, β
(b) β, α
(c) α, $(1 - \beta)$
(d) $(1 - \alpha)$, β
(e) $(1 - \alpha)$, $(1 - \beta)$

Answer: (c)
Difficulty level: Easy to Medium

9.3.9: You *cannot* make a type II error when:

(a)   the null hypothesis is true
(b)   the null hypothesis is false
(c)   the significance level is greater than .05
(d)   the significance level is less than .05
(e)   type II errors are always possible, regardless of our
      decision to accept or reject the null hypothesis

Answer: (a)
Difficulty level: Easy to Medium

9.3.10: The probability of type I and type II errors *must*:

(a)   be equal
(b)   sum to 1
(c)   be inversely related
(d)   both decrease as the sample size increases
(e)   both increase as the sample size increases

Answer: (c)
Difficulty level: Easy to Medium

9.3.11: Which significance level (α) would place the
greatest burden of proof on the alternative hypothesis?

(a)   10% for a one-sided test
(b)   10% for a two-sided test
(c)   5% for a one-sided test
(d)   1% for a one-sided test
(e)   1% for a two-sided test

Answer: (e)
Difficulty level: Easy to Medium

9.4.1: Which of the following *is not* one of the steps in hypothesis testing?

(a)   formulating the null and alternative hypotheses
(b)   selecting the sample size n and setting α
(c)   collecting the data and calculating the test statistic
(d)   comparing calculated and critical values in order to make a decision to accept or reject the null hypothesis
(e)   all of these are steps in hypothesis testing

Answer: (e)
Difficulty level: Easy

9.4.2: *Statistical significance*:

(a)   implies that business decisions should not depend on the economic and practical implications of the decision
(b)   implies that H_0 can be rejected for any level of α
(c)   implies that H_0 cannot be rejected for any level of α
(d)   does not always imply practical significance
(e)   none of these

Answer: (d)
Difficulty level: Easy to Medium

9.5.1: What is(are) the critical value(s) at $\alpha = .05$ for a test of H_0: $\mu \geq 100$ if the population is *normal* with a known standard deviation of 30?   (Assume n = 25 and $\alpha = .05$.)

(a)   111.76
(b)   90.13
(c)   88.24
(d)   109.87
(e)   The critical value cannot be determined since n < 30

Answer: (b)
Difficulty level: Easy to Medium

9.5.2: What is(are) the critical value(s) at α = .05 for a test of H_0: μ = 100 if the population is *normal* with an unknown population variance? (Assume n = 9 and s = 24.)

(a)   114.88
(b)   85.12
(c)   85.12 and 114.88
(d)   81.55 and 118.45
(e)   84.32 and 115.68

Answer: (d)
Difficulty level: Easy to Medium

9.5.3: What is(are) the critical value(s) at α = .05 for a test of H_0: $\mu \leq$ 100 if the population is *normal* with an unknown population variance? (Assume n = 9 and s = 24).

(a)   113.16
(b)   114.88
(c)   118.45
(d)   81.55
(e)   81.55 and 118.45

Answer: (b)
Difficulty level: Easy to Medium

9.5.4: Given $\overline{X}$ = 100, n = 10, and a *normal* population with an unknown standard deviation, what is(are) the critical Z (or t) value(s) when α = .01 and H_1: $\mu <$ 110?

(a)   t = ± 2.575
(b) $t_{.01,9}$ = -2.821
(c) $t_{.005,9}$ = ± 3.250
(d) $Z_{.01}$ = -2.33
(e)   Cannot be determined from the information given

Answer: (b)
Difficulty level: Easy to Medium

9.5.5: Given $\overline{X} = 100$, n = 10, s = 20, and a *normal* population with an unknown standard deviation, what is(are) the critical Z (or t) value(s) when $\alpha = .01$ and H_a: $\mu \neq 110$?

(a) t = -1.58
(b) $t_{.01,9} = -2.821$
(c) $t_{.005,9} = \pm 3.250$
(d) $Z_{.01} = \pm 2.575$
(e) Cannot be determined from the information given

Answer: (c)
Difficulty level: Easy to Medium

9.5.6: Given $\overline{X} = 100$, n = 10, s = 20, and a population that is *normal* with an unknown standard deviation, what is the value of the sample statistic when $\alpha = .01$ and H_a: $\mu \neq 110$?

(a) t = -1.58
(b) Z = -1.58
(c) $t_{.01,9} = -2.821$
(d) $t_{.005,9} = \pm 3.250$
(e) $Z_{.005} = \pm 2.575$

Answer: (a)
Difficulty level: Easy to Medium

9.5.7: Given $\overline{X} = 100$ and s = 20, obtained from a random sample of n = 10 items without replacement from a finite *normal* population of N = 100 items, what is the value of the sample statistic for testing the hypothesis H_a: $\mu \neq 110$?

(a) t = -1.58
(b) t = -1.66
(c) $t_{.01,9} = -2.821$
(d) $t_{.005,9} = \pm 3.250$
(e) Cannot be determined

Answer: (b)
Difficulty level: Medium

9.5.8: Given a *normal* population and the information below, at the given level of significance, our decision is:

$n = 25$, $\overline{X} = 28$, $s = 20$, H_0: $\mu \leq 20$, H_a: $\mu > 20$, $\alpha = .05$

(a) Reject H_0 and conclude that $\mu \leq 20$
(b) Reject H_0 and conclude that $\mu > 20$
(c) Reject H_0 and conclude that $\mu \neq 20$
(d) Do not reject H_0 and conclude that $\mu > 20$
(e) Do not reject H_0 and conclude that we do not have strong sample evidence against the null hypothesis

Answer: (b)
Difficulty level: Easy to Medium

9.5.9: Given a *normal* population and the information below, at the given level of significance, our decision is:

$n = 25$, $\overline{X} = 14$, $s = 15$, H_0: $\mu = 20$, H_a: $\mu \neq 20$, $\alpha = .05$

(a) Do not reject H_0 and conclude that we do not have strong sample evidence against the null hypothesis
(b) Do not reject H_0 and conclude that $\mu \neq 20$
(c) Reject H_0 and conclude that $\mu = 20$
(d) Reject H_0 and conclude that $\mu > 20$
(e) Reject H_0 and conclude that $\mu \neq 20$

Answer: (a)
Difficulty level: Medium

9.6.1: Given a *normal* population and the information below, at the given level of significance, our decision is:

$n = 25$, $\overline{X} = 14$, $\sigma = 15$, H_0: $\mu = 20$, H_a: $\mu \neq 20$, $\alpha = .05$

(a) Do not reject H_0, since the P-value is less than .05
(b) Do not reject H_0, since the P-value exceeds .05
(c) Do not reject H_0, since the P-value exceeds .025
(d) Reject H_0, since the P-value is less than .025
(e) Reject H_0, since the P-value is less than .05

Answer: (e)
Difficulty level: Medium

9.6.2: Given a *normal* population and the information below, at the given level of significance, our decision is:

$n = 9$, $\overline{X} = 50$, $s^2 = 100$, H_0: $\mu = 55$, H_a: $\mu \neq 20$, $\alpha = .05$

(a) Do not reject H_0, since the P-value is less than .10
(b) Do not reject H_0, since the P-value exceeds .05
(c) Do not reject H_0, since the P-value exceeds .10
(d) Reject H_0, since the P-value exceeds .05
(e) Reject H_0, since the P-value is less than .05

Answer: (b)
Difficulty level: Medium

9.6.3: In a sample of 400 bushings manufactured by Algonquin Corporation there were 12 bushings whose diameters were not within the specified acceptable tolerances. At $\alpha = .05$, the P-value for testing whether H_0: $p \leq .02$ is:

(a) .1528
(b) 1.43
(c) .4236
(d) .0764
(e) Cannot be determined without further information

Answer: (d)
Difficulty level: Easy to Medium

9.6.4: A *P-value* is a (the) probability:

(a) of committing a Type I error
(b) of committing a Type I error
(c) that the test statistic would assume a value as or more extreme than the critical value
(d) that the test statistic would assume a value as or more extreme than the observed value of the test statistic
(e) that is the same for both one-sided and two-sided tests

Answer: (d)
Difficulty level: Easy to Medium

9.7.1: What is(are) the critical value(s) for a test of the null hypothesis H_0: p = .100, given n = 100 independent trials from a *binomial* population, and α = .05?

(a)   .0412
(b)   .1588
(c)   .0412 and .1588
(d)   .0506 and .1494
(e)   Cannot be determined

Answer: (c)
Difficulty level: Easy to Medium

9.7.2: At α = .01, what is(are) the critical Z (or t) value(s) for a test of the null hypothesis H_0: p = .100, given n = 16 independent trials from a *binomial* population?

(a) $t_{.01,15}$ = ± 2.602
(b) $t_{.005,15}$ = ± 2.947
(c) $Z_{.005}$ = ± 2.575
(d) $Z_{.01}$ = ± 2.33
(e)   Cannot be determined since np(1 - p) < 5

Answer: (c)
Difficulty level: Medium

9.7.3: What is the value of the test statistic, given a *binomial* population, H_0: p = .100, n = 16, and α = .02?

(a)   .100 ± .075
(b)   .100 ± (2.33)(.075)
(c)   .100 ± (2.17)(.075)
(d)   .100 ± (2.33)(.0056)
(e)   Cannot be determined since np(1 - p) < 5

Answer: (e)
Difficulty level: Medium

9.7.4: A sample of n = 144 registered voters from a *binomial* population found that 84 were in favor of a bond issue. If testing the hypothesis H_a: $p \neq .50$, our decision at $\alpha = .10$ should be:

(a)   do not reject H_0 and conclude that $p \neq .50$
(b)   do not reject H_0 and conclude that $p > .50$
(c)   reject H_0 and conclude that $p = .50$
(d)   reject H_0 and conclude that $p \neq .50$
(e)   reject H_0 and conclude that $\mu \neq 50$

Answer: (d)
Difficulty level: Medium

9.8.1: In a two population problem, if both populations are *not* normally distributed and both population standard deviations are unknown, then the t score:

(a)   is not appropriate
(b)   is appropriate if and only if $(n_1 + n_2 - 2) \geq 30$
(c)   is appropriate if and only if $n_1 \geq 30$ and $n_2 \geq 30$
(d)   is appropriate if and only if samples are random and independent
(e)   is appropriate if and only if we pool the sample variances and $\sigma_1 = \sigma_2 = \sigma$

Answer: (a)
Difficulty level: Medium

9.8.2: If two populations are normally distributed, the sampling distribution for the difference in sample means will be:

(a)   normally distributed
(b)   normally distributed if and only if $(n_1 + n_2 - 2) \geq 30$
(c)   normally distributed if and only if $n_1 \geq 30$ and $n_2 \geq 30$
(d)   normally distributed if and only if samples are random and independent
(e)   normally distributed if and only if we σ_1 and σ_2 are both known and equal to one another

Answer: (a)
Difficulty level: Easy to Medium

9.8.3: The appropriate test statistic to use when testing the hypothesis H_0: $\mu_1 = \mu_2$ when both populations are normally distributed and σ_1 and σ_2 are both known, is:

(a)   t, with $(n_1 + n_2 - 1)$ degrees of freedom
(b)   t, with $(n_1 + n_2 - 2)$ degrees of freedom
(c)   Z, with $(n_1 + n_2 - 2)$ degrees of freedom
(d)   Z, the standard normal random variable
(e)   Chi-square, with $(n_1 - 1)(n_2 - 1)$ degrees of freedom

Answer: (d)
Difficulty level: Easy

9.8.4: The appropriate test statistic to use when testing hypotheses for the *difference in population means*, when both populations are normally distributed and σ_1 and σ_2 are both unknown, but assumed to be equal, is:

(a)   t, with $(n_1 + n_2 - 1)$ degrees of freedom
(b)   t, with $(n_1 + n_2 - 2)$ degrees of freedom
(c)   Z, with $(n_1 + n_2 - 2)$ degrees of freedom
(d)   Z, the standard normal random variable
(e)   Chi-square, with $(n_1 - 1)(n_2 - 1)$ degrees of freedom

Answer: (b)
Difficulty level: Easy

9.8.5: The variance of the sampling distribution for the difference in sample means will equal $[\sigma_1^2/n_1 + \sigma_2^2/n_2]$ provided that:

(a)   samples are random and independent of one another
(b)   both populations are normally distributed
(c)   both samples are sufficiently large (at least 30)
(d)   both population variances are known and equal
(e)   we pool the sample variances

Answer: (a)
Difficulty level: Medium

CHAPTER 9

9.8.6: Which of the following conditions or assumptions is necessary for us to be able to test $H_0: \mu_1 - \mu_2 = \delta_0$?

(a) Both samples are independent random samples
(b) Both populations are normal
(c) Both samples are sufficiently large (at least 30)
(d) Both population standard deviations are known and equal
(e) All of these must be true to test the above hypothesis

Answer: (a)
Difficulty level: Medium

9.8.7: Excess returns for random samples of two different investments are listed below. For testing the hypothesis $H_0: \mu_1 = \mu_2$ in the case where both populations are normal and the population variances are unknown but assumed to be equal, the degrees of freedom are:

Sample	n_j	$\overline{X}_j$	s_j
1	8	10	4
2	9	6	2

(a) 7.88
(b) 10
(c) 11
(d) 15
(e) 16

Answer: (b)
Difficulty level: Medium

9.8.8: Independent random samples from normal distributions with equal standard deviations of excess percentage returns during similar economic conditions for stocks (1) just after and (2) just before stock split announcements. The results are listed below. Given $H_0: \mu_1 - \mu_2 \le 5$, $\alpha = .01$, we should:

Sample	n_j	$\overline{X}_j$	s_j
1	8	10	4
2	9	6	2

(a) Do not reject H_0, as the test statistic is below 2.602
(b) Reject H_0, since the test statistic is less than 2.602
(c) Do not reject H_0, since the P-value is less than .01
(d) Reject H_0, since the P-value is less than .01
(e) Continue sampling since the results are inconclusive

Answer: (a)
Difficulty level: Medium

9.9.1: For a *difference in proportions* hypothesis test, the standard normal or Z distribution is used to standardize values of the difference in sample proportions:

(a) if the conditions $np \geq 5$ and $n(1 - p) \geq 5$ are satisfied
(b) if the condition $(n_1 + n_2 - 2) \geq 30$ is satisfied
(c) if the conditions $n_1 \geq 30$ and $n_2 \geq 30$ are both satisfied
(d) if the conditions $np < 5$ and $n(1 - p) < 5$ are satisfied
(e) instead of t because the populations are not normal

Answer: (d)
Difficulty level: Easy to Medium

9.9.2: For a *difference in proportions* hypothesis test, the two estimates of the population proportion should be pooled whenever:

(a) the null hypothesis is that the difference in population proportions is equal to zero.
(b) the alternative hypothesis is that the difference in population proportions is greater than zero.
(c) the alternative hypothesis is that the difference in population proportions is less than zero.
(d) the statistician is trying to determine whether there is a significant (positive or negative) difference between two population proportions.
(e) the two estimates of the population proportion should be pooled in each of the above situations.

Answer: (e)
Difficulty level: Easy to Medium

9.9.3: A brick-making firm makes bricks by two different processes. In samples of 200 bricks from the first process and 300 bricks from the second process, it was found that 24 of the first type broke during baking in the kiln and 51 of the second type broke in the kiln. If we test the hypothesis H_0: $p_1 = p_2$, we should (at $\alpha = .10$):

(a) Do not reject H_0, and conclude that the sample evidence is not strong enough to reject the null hypothesis.
(b) Reject H_0, and conclude that the sample evidence is strong enough to reject the null hypothesis.
(c) Do not reject H_0, and conclude that $p_1 \neq p_2$.
(d) Reject H_0, and conclude that $p_1 = p_2$.
(e) Conclude that the results are indeterminate.

Answer: (a)
Difficulty level: Easy to Medium

9.9.4: Random samples of 200 men and 100 women with MBA degrees revealed that 120 men and only 45 women received promotions within two years on the job. We would like to know whether the percentage of women with MBA degrees who receive promotions (p_{women}) is *less than* the proportion of men with MBA degrees who receive promotions (p_{men}). At $\alpha = .01$, we should:

(a)   Do not reject H_0 and conclude that $p_{women} \geq p_{men}$
(b)   Do not reject H_0 and conclude that $p_{women} \neq p_{men}$
(c)   Reject H_0 and conclude that $p_{women} < p_{men}$
(d)   Reject H_0 and conclude that $p_{women} \neq p_{men}$
(e)   Reject H_0 and conclude that $p_{women} < p_{men}$

Answer: (e)
Difficulty level: Easy to Medium

9.10.1: Which of the assumptions below are necessary for the test statistic of hypothesis tests used to test whether H_0: $\sigma_1^2 = \sigma_2^2$ to have an F distribution that is well-defined?

I.   samples are random and independent of one another
II.   the populations are both normally distributed
III. the population standard deviations are equal

(a)   I only
(b)   II only
(c)   I and II only
(d)   I and III only
(e)   I, II, and III are all necessary for the test statistic to follow an F distribution and be well-defined

Answer: (c)
Difficulty level: Easy to Medium

9.10.2: If the calculated value for the F test to determine whether two populations have equal variances is very large, then:

(a)   it is highly likely that the two populations have equal variances, since the P-value is very large
(b)   it is highly unlikely that the two populations have equal variances since the P-value is very large
(c)   it is highly likely that the two populations have equal variances, since the P-value is very small
(d)   it is highly unlikely that the two populations have equal variances, since the P-value is very small
(e)   we must accept the null hypothesis that the two populations have equal variances

Answer: (d)
Difficulty level: Medium

9.10.3: Independent random samples of size 10 are selected from each of two populations. The sample means are calculated to be 10 and 12.5, respectively. The sample standard deviations are also calculated and these values are 2 and 4, respectively. If we test the hypothesis that these two populations have equal variances at $\alpha = .05$, we should:

(a)   do not reject H_0, since the P-value exceeds .05
(b)   do not reject H_0, since the P-value is less than .05
(c)   reject H_0, since the P-value exceeds .05
(d)   reject H_0, since the P-value is less than .05
(e)   start over, since the test results are inconclusive

Answer: (a)
Difficulty level: Medium

9.11.1: Other things equal, a researcher is more likely to:

(a)   reject the null with a one-sided test than with a two-sided test
(b)   reject the null with a two-sided test than with a one-sided test
(c)   reduce sampling error by reducing sampling size
(d)   more than one of these is correct
(e)   accept the null hypothesis with a one-sided test than with a two-sided test

Answer: (a)
Difficulty level: Easy to Medium

9.11.2: Which of the following statements are true?

(a)   The assumption of random samples can be ignored if we make probability statements in connection with our conclusion.
(b)   The assumption of normality is a strong assumption because of the central limit theorem effect.
(c)   If the standard deviations are equal and we use methods assuming that they are unequal, the test becomes less conservative.
(d)   As the sample size n increases, the standard normal distribution tends to closely approximate the t distribution.
(e)   None of these statements are true.

Answer: (d)
Difficulty level: Easy to Medium

9.11.3: Corn Oil, Inc., produces 1-gallon jugs of oil that are reported to contain 128 ounces of oil. The value of σ is known to be 4 ounces. A sample of 300 jugs is used to test the claim that the jugs contains at least 128 ounces. This sample resulted in a mean content of 125 ounces. At $\alpha = .01$, is the above claim valid?

(a)   Yes, failure to reject H_0 is strong evidence that the claim is valid
(b)   Yes, failure to reject H_0 is relatively weak evidence that the claim is valid
(c)   No, failure to reject H_0 is strong evidence that the claim is invalid
(d)   No, rejection of H_0 is relatively weak evidence that the claim is invalid
(e)   We cannot say for sure as the results are indeterminate

Answer: (d)
Difficulty level: Easy to Medium

9.11.4: A random sample of 200 voters is selected from all
registered voters in a certain city, and 175 of these voters
indicated that they planned to vote for Mr. Green for mayor.
Mr. Green had recently claimed that he expected to get at
least 90% of the vote. Answer the following using $\alpha = .05$.

a.   What are the appropriate null and alternative
      hypotheses to test Mr. Green's claim?
b.   What are the calculated and critical Z (or t) values?
c.   For the above test, what inference should we make?
d.   Would the inference made in part c have been different
      if we had instead used $\alpha = .01$? $\alpha = .10$?

Answer: a. H_0: $p \geq .90$; H_a: $p < .90$

           b. $Z = -1.18$, $-Z_{.05} = -1.645$
           c.   Do not reject H_0, since Z is smaller in
                 absolute value than 1.645
           d.   No
Difficulty level: Easy to Medium

9.11.5: A University professor teaches two sections of the
same course, one at 8:00 am and one at 9:00 am. Recently he
gave the same thirty-point quiz to both sections. He
worried that the students in the second section might have
received information from students in the first section,
thereby improving their grades. The results are listed
below. Answer the following using $\alpha = .01$.

8:00 am Section	**9:00 am Section**
$n_1 = 62$	$n_2 = 87$
$\overline{X}_1 = 16.03$	$\overline{X}_2 = 16.28$
$s_1 = 5.05$	$s_2 = 5.29$

a.   What are the appropriate null and alternative
      hypotheses to investigate the professor's worries?
b.   What are the calculated and critical Z (or t) values?
c.   For the above test, what inference should we make?
d.   Would the inference made in part c have been different
      if we had instead used $\alpha = .05$? $\alpha = .10$?

Answer: a. H_0: $\mu_1 \geq \mu_2$; H_a: $\mu_1 < \mu_2$

           b. $Z = -.29$, $-Z_{.01} = -2.33$
           c.   Do not reject H_0, since Z is smaller in
                 absolute value than 2.33
           d.   No
Difficulty level: Easy to Medium

9.11.6: Hi-note Company produces a line of notebooks for the college market. Notebook line C consists of 100 pages and the manufacturer regularly draws samples of size 25 books to test if they deviate from the 100 pages. One sample yielded a mean of 95 pages and a standard deviation of s = 10 pages. <u>Answer the following using α = .05</u>.

a.   What are the appropriate null and alternative hypotheses for testing the implicit claim above?
b.   What are the calculated and critical Z (or t) values?
c.   For the above test, what inference should we make?
d.   Would the inference made in part c have changed if we had instead used α = .01? α = .10?

Answer: a. $H_0: \mu = 100$; $H_a: \mu \neq 100$

          b. $t = -2.50$, $t_{.05,24} = 1.711$
          c.   Reject H_0, since t is larger in absolute value than 1.711. Conclude that $\mu \neq 100$ pages
          d.   No
Difficulty level: Easy to Medium

9.11.7: Two marketing companies were asked to try selling a new product. Each company randomly selected and approached 400 customers. The first company had 80 successful sales while the second company had 50 successful sales. After seeing these results, the president of the second company said "Why, our sales success rate is as good or better than the other company's in the long run. We were just a little behind in this test." <u>Answer the following using α = .05</u>.

a.   What are the appropriate null and alternative hypotheses for testing the claim made by the president of the second company?
b.   What are the calculated and critical Z (or t) values?
c.   For the above test, what inference should we make?
d.   Would the inference made in part c have changed if we had instead used α = .01? α = .10?

Answer: a. $H_0: p_1 \leq p_2$; $H_a: p_1 > p_2$

          b. $Z = 2.875$, $Z_{.05} = 1.645$
          c.   At α = .05, we reject H_0 and conclude that $p_1 > p_2$. We reject company 2 president's claim that their sales success rate is at least as high as that of company 1
          d.   No
Difficulty level: Easy to Medium

9.11.8: A new computer company claims that the mean time between failure of its hard disk drives exceeds that of the IBM-PC; hence, its microcomputer is more reliable than the IBM-PC. Assume that both populations are normally distributed with equal population variances. The results are listed below. <u>Answer the following using $\alpha = .05$.</u>

New Company	IBM-PC
$n_1 = 5$	$n_2 = 5$
$\overline{X}_1 = 16,000$	$\overline{X}_2 = 12,000$
$s_1 = 3,000$	$s_2 = 2,000$

a.   What are the appropriate null and alternative hypotheses for testing the above claim?
b.   What are the calculated and critical Z (or t) values?
c.   For the above test, what inference should we make?

Answer: a. $H_0: \mu_{new} \leq \mu_{IBM}; H_a: \mu_{new} > \mu_{IBM}$

b. $t = 1.94$, $t_{.05,8} = 1.86$
c.   Reject H_0, and conclude that the new microcomputer is more reliable than the IBM-PC.
Difficulty level: Easy to Medium

9.11.9: A light bulb manufacturer is considering producing a new type of light. The manufacturer has decided that she will only produce the new bulb only if she is convinced that its average life *exceeds* that of the old bulb. Random samples of 20 old and 25 new bulbs yield sample means of 11,500 and 12,500 hours and sample standard deviations of 1,500 and 2,000 hours. <u>Answer the following using $\alpha = .05$.</u>

a.   What are the appropriate null and alternative hypotheses the manufacturer should use to decide whether to produce the new light bulb?
b.   What are the calculated and critical Z (or t) values?
c.   For the above test, what inference should we make?
d.   Would the inference made in part c have changed if we changed whether the test was one-sided or two-sided?

Answer: a. $H_0: \mu_{old} \geq \mu_{new}; H_a: \mu_{old} < \mu_{new}$

b. $Z = -1.86$, $-Z_{.05} = -1.645$

c.   At $\alpha = .05$, she should reject H_0 and go ahead and produce the new light bulb.
d.   Yes, since $Z_{.025} = 1.96$.
Difficulty level: Easy to Medium

9.12.1: Which of the following would increase the *power* of a given hypothesis test (all else held constant)?

I.   increasing the sample size n
II. increasing the level of significance α

(a) only I would increase the power of the test
(b) only II would increase the power of the test
(c) both I and II would increase the power of the test
(d) neither I nor II would increase the power of the test
(e) neither I nor II would change the power of the test

Answer: (c)
Difficulty level: Medium

9.12.2: Holding the sample size constant, the probability of making a Type II error can only be reduced by:

(a) decreasing the width of the rejection region
(b) increasing the width of the acceptance region
(c) reducing the power of the test
(d) increasing the probability of making a Type I error
(e) decreasing the probability of making a Type I error

Answer: (d)
Difficulty level: Medium

9.12.3: Given a population that is normally distributed with a mean of $\mu = 25$ and a standard deviation of $\sigma = 5$, the *power* of the test of the null hypothesis that the mean is less than or equal to 24, given $\alpha = .05$ and n = 100, is:

(a) .36
(b) .64
(c) .50
(d) .69
(e) 1.00

Answer: (b)
Difficulty level: Medium

9.13.1: A company manager would like to determine whether a new machine is more efficient than the one currently being used. He records the output of n = 16 workers on both the current used and new machines. These two samples are:

(a)   independent
(b)   paired or dependent
(c)   mutually exclusive
(d)   random
(e)   stratified

Answer: (b)
Difficulty level: Easy

9.13.2: You are given the following paired or dependent samples: X_1 = [4, 10, 21, 37], X_2 = [7, 11, 22, 38]. Given that both populations are normally distributed, the critical Z (or t) values for testing H_0: $\mu_1 - \mu_2 = 0$ at α = .05 are _____ and the appropriate decision is to _____.

(a)   ± 1.96; reject the null hypothesis
(b)   ± 2.353; do not reject the null hypothesis
(c)   ± 2.353; reject the null hypothesis
(d)   ± 3.182; do not reject the null hypothesis
(e)   ± 3.182; reject the null hypothesis

Answer: (d)
Difficulty level: Easy to Medium

9.13.3: Twenty housewives are each given two boxes of laundry detergent, detergent A and detergent B, and are asked to use the two detergents on sets of white laundry. Each housewife then brings in the cleaned laundry to be checked for its' whiteness. The appropriate test is a(n):

(a)   F test
(b)   difference in means test
(c)   difference in proportions test
(d)   test for the equality of population variances
(e)   paired difference in means test

Answer: (e)
Difficulty level: Easy to Medium

Analysis of Variance

10.1.1: Which of the following is *not* an assumption required for analysis of variance?

(a) samples are random and independent of one another
(b) each of the populations are normally distributed
(c) each of the populations have equal variances
(d) each of the population variances are known
(e) all of these assumptions are equally important

Answer: (a)
Difficulty level: Easy to Medium

10.1.2: The correct formulation for the null hypothesis in an analysis of variance (ANOVA) situation would be:

(a) $\mu_1 = \mu_2$

(b) $\mu_1 = \mu_2 = \ldots = \mu_k$

(c) $\mu_1 = \mu_2 = \ldots = \mu_k = 0$

(d) $\mu_1 + \mu_2 + \ldots + \mu_k = 1$

(e) $\mu_1 + \mu_2 + \ldots + \mu_k = 0$

Answer: (b)
Difficulty level: Easy

10.1.3: Random and independent samples were taken from three normally distributed populations with equal variances to test whether there is any significant difference in the mean number of units being produced weekly by each of the three production methods, with the results listed below. At $\alpha = .05$, the within-group and between-group variance estimates:

$$\overline{X}_1 = 10, \qquad \overline{X}_2 = 12, \qquad \overline{X}_3 = 14$$

$$s_1 = 2, \qquad s_2 = 3, \qquad s_3 = 3$$

$$n_1 = 5, \qquad n_2 = 4, \qquad n_3 = 6$$

(a) are 7.33 and 22
(b) are 22 and 7.33
(c) are 4.091 and 3.8056
(d) are 4.091 and 4.9653
(e) cannot be determined using the information above

Answer: (a)
Difficulty level: Easy to Medium

10.1.4: Random and independent samples were taken from three normally distributed populations with equal variances to test whether there is any significant difference in the mean number of units being produced weekly by each of the three production methods, with the results listed below. At α = .05, the calculated and critical F values for testing the hypothesis that the three populations have equal means:

$$\overline{X}_1 = 10, \qquad \overline{X}_2 = 12, \qquad \overline{X}_3 = 14$$

$$s_1 = 2, \qquad s_2 = 3, \qquad s_3 = 3$$

$$n_1 = 5, \qquad n_2 = 4, \qquad n_3 = 6$$

(a)   are 7.33 and 22
(b)   are 22 and 7.33
(c)   are 3.001 and 3.89
(d)   are 3.001 and 3.81
(e)   cannot be determined using the information above

Answer: (c)
Difficulty level: Easy to Medium

10.2.1: Given the assumption that each of the k populations have equal variances, the _____ will provide an unbiased estimate of the common population variance.

(a)   between-group sum of squares
(b)   within-group sum of squares
(c)   between-group sum of squares divided by (k - 1)
(d)   within-group sum of squares divided by (n_T - k)
(e)   ratio of the numerator of the F statistic to the denominator of the F statistic

Answer: (d)
Difficulty level: Easy to Medium

10.2.2: The numerator of the F statistic is an estimate of the unknown common variance that is based on:

(a)   the assumption that all of the populations are normal
(b)   the assumption that samples are random and independent
(c)   the within-groups variance estimate
(d)   the between-groups variance estimate
(e)   the total sum of squares

Answer: (d)
Difficulty level: Easy to Medium

10.2.3: The critical F value for a test of the hypothesis that three population means are equal depends on each of the following *except*:

(a)   the value chosen for α
(b)   the degrees of freedom in the numerator
(c)   the degrees of freedom in the denominator
(d)   the ratio of the between-group and within-group variance estimates; hence, $F = s^2_B/s^2_W$
(e)   the critical F value depends on all of these

Answer: (d)
Difficulty level: Easy

10.2.4: The F distribution is:

(a)   the ratio of two chi-square random variables
(b)   the ratio of two student t random variables
(c)   the ratio of two standard normal or Z random variables
(d)   the ratio of two chi-square random variables, each of which being divided by its degrees of freedom
(e)   a symmetric and bell-shaped continuous distribution

Answer: (d)
Difficulty level: Easy

10.2.5: In analysis of variance, the farther apart the sample means are from one another, the more likely:

(a)   the null hypothesis will be accepted
(b)   the null hypothesis will be rejected
(c)   the F statistic will be less than one
(d)   the populations are normally distributed
(e)   the sample variances will differ from one another

Answer: (b)
Difficulty level: Easy

10.2.6: Different teaching methods are used experimentally in five sections of introductory statistics. If we sample six student scores on a recent exam from each section to test whether the teaching methods made any difference, the appropriate test statistic would be:

(a)   Chi-square, with 9 degrees of freedom
(b)   Chi-square, with 20 degrees of freedom
(c)   F, with 20 degrees of freedom
(d)   F, with 4,25 degrees of freedom
(e)   F, with 4,5 degrees of freedom

Answer: (d)
Difficulty level: Easy to Medium

10.2.7: Random and independent samples are taken from three normal populations with equal variances. The results are: A = [8, 10, 12], B = [7, 9, 11], and C = [13, 10, 13]. The grand or overall mean is:

(a)   9
(b)   10
(c)   11
(d)   12
(e)   10.33

Answer: (e)
Difficulty level: Easy

10.2.8: Random and independent samples are taken from three normal populations with equal variances. The results are: A = [8, 10, 12], B = [7, 9, 11], and C = [13, 10, 13]. The within-group and between-group variance estimates are:

(a)   7 and 3.67
(b)   7 and 22
(c)   14 and 3.67
(d)   14 and 22
(e)   3.67 and 7

Answer: (a)
Difficulty level: Easy to Medium

10.2.9: Random and independent samples are taken from three normal populations with equal variances. The results are: A = [8, 10, 12]. B = [7, 9, 11], and C = [13, 10, 13]. The calculated and critical F values for the test of the null hypothesis that the three populations have the same mean, at α = .01, are:

(a)   1.91 and 14.544
(b)   .636 and 14.544
(c)   1.91 and 10.925
(d)   .636 and 10.925
(e)   10.925 and .636

Answer: (c)
Difficulty level: Medium

10.2.10: Random and independent samples are taken from three normal populations with equal variances. The results are: A = [8, 10, 12], B = [7, 9, 11], and C = [13, 10, 13]. The appropriate decision, at α = .05, is to:

(a)   do not reject H_0: $\mu_1 = \mu_2 = \mu_3$, since F < 5.1433
(b)   do not reject H_0: $\mu_1 = \mu_2 = \mu_3$, since F < 7.2598
(c)   reject H_0: $\mu_1 = \mu_2 = \mu_3$, since F < 5.1433
(d)   reject H_0: $\mu_1 = \mu_2 = \mu_3$, since F > 1.645
(e)   perform the test again using other random samples

Answer: (a)
Difficulty level: Medium

10.2.11: Given the following ANOVA computer printout, the between-group and within-group variance estimates are:

Source	df	SS	Variances or Mean Squares	F
Between groups	3	348	116	11.6
Within groups	8	80	10	
Total	11	428		

(a)   348 and 80
(b)   80 and 348
(c)   116 and 10
(d)   10 and 116
(e)   none of these

Answer: (c)
Difficulty level: Easy to Medium

10.2.12: Given the following ANOVA computer printout, the appropriate decision, at $\alpha = .01$, is to:

Source	df	SS	Variances or Mean Squares	F
Between groups	3	348	116	11.6
Within groups	8	80	10	
Total	11	428		

(a) do not reject H_0, since F > 7.591
(b) do not reject H_0, since F > 9.5965
(c) reject H_0, since F > 7.591
(d) reject H_0, since F > 9.5965
(e) perform the test again using other random samples

Answer: (c)
Difficulty level: Easy to Medium

10.2.13: Random and independent samples are taken from three normal populations with equal variances. The results are: A = [6, 10, 14], B = [7, 8, 9], and C = [13, 11]. The between-group and within-group variance estimates, are:

(a) 9.75 and 5.14
(b) 5.14 and 9.75
(c) 2.79 and 5.14
(d) 3 and 8
(e) 3 and 9.75

Answer: (a)
Difficulty level: Easy to Medium

10.2.14: Random and independent samples are taken from three normal populations with equal variances. The results are: A = [6, 10, 14], B = [7, 8, 9], and C = [13, 11]. The number of treatments and total number of observations are:

(a) 20 and 20
(b) 10 and 20
(c) 3 and 8
(d) 10 and 4
(e) 2 and 5

Answer: (c)
Difficulty level: Easy

10.2.15: Random and independent samples are taken from three normal populations with equal variances. The results are: A = [6, 10, 14], B = [7, 8, 9], and C = [13, 11]. At α = .05, we _____ the null hypothesis that the three populations have the same mean because _____.

(a) cannot reject; F is less than 5.7861
(b) cannot reject; F exceeds 8.4336
(c) should reject; F is less than 8.4336
(d) should reject; F exceeds 5.7861
(e) should reject; F exceeds 1.90

Answer: (a)
Difficulty level: Medium

10.2.16: An engineer was interested in knowing the wear characteristic for three different brands of steel-belted radial tires. Random and independent samples of *ten* tires for each brand were taken, with the results (reported in thousands of miles) listed in the table below. The engineer is interested in knowing whether the three brands of tires have the same mean, assuming equal variances for each population. Answer the following using α = .05.

	Brand A	Brand B	Brand C
Average mileage	36.40	38.20	33.10
Sample standard deviation	1.65	1.80	1.50

a. What are the appropriate null and alternative hypotheses for testing whether the three tires have the same mean wear characteristics?
b. What are the between-group and within-group variance estimates?
c. What inference should we make?

Answer: a. H_0: $\mu_1 = \mu_2 = \mu_3$
             H_1: At least one of the means is different
         b. s^2_B = 66.90, s^2_W = 2.7375
         c. We should reject H_0 and conclude that the three brands of tires have different means
Difficulty level: Medium

10.2.17: A manager is interested in determining whether the productivity of workers that work during three different shifts (day, evening, and graveyard) is the same. To test this hypothesis, the manager randomly samples eight workers from each shift and records the average time needed to complete a given assembly-line task. Assuming that the populations are normal with equal variances:

(a)   the appropriate test is a chi-square test
(b)   degrees of freedom for the denominator are 18
(c)   degrees of freedom for the numerator are 3
(d)   the calculated F value must exceed the critical value
(e)   none of these

Answer: (e)
Difficulty level: Easy to Medium

10.2.18: A manager is interested in determining whether the productivity of workers that work during three different shifts (day, evening, and graveyard) is the same. To test this hypothesis, the manager randomly samples eight workers from each shift and records the average time (in minutes) needed to complete a given assembly-line task, with the results summarized in the table below. Assume that the populations are normal with equal variances. At a significance level of $\alpha = .01$, we should:

	Day shift	Night shift	Graveyard shift
n_j	8	8	8
$\overline{X}_j$	4.90	6.20	7.50
s_j	1.40	.80	1.10

(a)   do not reject H_0 and conclude that the three population mean times to complete an assembly-line task are equal
(b)   do not reject H_0 and conclude that the three population mean times to complete an assembly-line task are significantly different from one another
(c)   reject H_0 and conclude that the three population mean times to complete an assembly-line task are equal
(d)   reject H_0 and conclude that at least one of the population mean times is different from the others
(e)   reject H_0 and conclude that all of the population mean times are significantly different from one another

Answer: (d)
Difficulty level: Medium

10.2.19: A national weight-loss clinic chain has been experimenting with three different diets. In order to test the effectiveness of the three diets, the research director took a random sample of forty-five women from across the country and offered them free clinic sessions if they would try the three diets. Thirty interested women went on the first diet for a month; then they went on the second diet for a month; and then on the third diet for a month. The weight losses are indicated below:

First Diet	Second Diet	Third Diet
$n_1 = 30$	$n_2 = 30$	$n_2 = 30$
$\overline{X}_1 = 10.2$	$\overline{X}_2 = 5.6$	$\overline{X}_3 = 2.2$
$s_1 = 3.0$	$s_2 = 3.3$	$s_3 = 3.0$

List the assumptions you have to make about the populations and/or the samples in order to perform analysis of variance. Do any of the assumptions appear to be violated in this situation? If any are, indicate which one(s) and how it (they) could be corrected.

Answer: The three assumptions are the following:
 a.   Samples are random and independent.
 b.   Populations are normally distributed.
 c.   All populations have the same variance.

The first assumption is badly violated. Randomness was avoided when the participants in the test were allowed to self-select themselves into the test. Independence of samples does not exist in this test since the same people were put on each diet. Anyone who diets knows that the first pounds come off more easily than those in the third month.

The normality of the populations is probably valid since the variable of interest is weight loss, and weights are often found to be normally distributed.

The equality of population variances seems to be a reasonable assumption since the smallest variance is 9.0 while the largest is 10.89. These are not much different. However, a better sampling design would involve the use of ninety women--thirty for each diet--selected randomly from among the new customers to the clinics during a set period of time. This way the samples would be independent, and the samples would be representative of the type of people who come to the clinic of their own accord.

Difficulty level: Easy to Challenging

10.2.20: Random and independent samples were taken from three normally distributed populations with equal variances to test whether there is any significant difference in the mean number of units being produced weekly by each of the three production methods, with the results listed below. The between-group and within-group degrees of freedom are:

$\overline{X}_1 = 10,$ $\overline{X}_2 = 12,$ $\overline{X}_3 = 14$

$s_1 = 2,$ $s_2 = 3,$ $s_3 = 3$

$n_1 = 5,$ $n_2 = 4,$ $n_3 = 6$

(a) 22 and 7.33
(b) 7.33 and 22
(c) 2 and 12
(d) 3 and 15
(e) unable to be determined with the information given

Answer: (c)
Difficulty level: Easy

10.3.1: In an analysis of variance problem, when all samples are *not* of equal size:

(a) the hypothesis test cannot be done
(b) a formula is used that allows one to adjust for different sample sizes
(c) some data must be eliminated, forcing all sample sizes to be equal
(d) the t-test should be used rather than the F-test
(e) all of these

Answer: (b)
Difficulty level: Easy

10.3.2: An information systems manager with ICX Corporation conducted a study of the operating time prior to failure for three brands of microcomputer hard disk drives. Random samples from normal populations with equal variances were selected. The results for the time to failure (in thousands of hours) are summarized in the table below. <u>Answer the following using</u> $\alpha = .01$.

	Brand A	Brand B	Brand C
n_j	6	5	4
$\overline{X}_j$	4.90	6.20	7.40
s_j	1.10	.500	.600

a. What is the grand mean for all of the sample data?
b. What are the between-group and within-group variance estimates?
c. What are the calculated and critical F values?
d. What inference should we make?

Answer: a. Grand Mean = 6.0

b. $s^2_B = 7.65$, $s^2_W = .6775$

c. $F = 7.65/.6775 = 11.29$, $F_{.01,2,12} = 6.9266$

d. Reject H_0; conclude that at least one of the population mean times to failure is differen

Difficulty level: Easy to Medium

10.4.1: Which of the following identities *must* be true in general when we partition the sum of squares?

(a) SST = SSR + SSE
(b) SST = SSB + SSW
(c) SSB = MSB/(k - 1)
(d) SSW = MSW/(n_T - k)
(e) more than one of these identities must be true in general when we partition the sum of squares

Answer: (b)
Difficulty level: Easy

10.4.2: When we divide a sum of squares by its associated degrees of freedom, we refer to the resulting value as:

(a)   the total sum of squares
(b)   the F statistic
(c)   a mean square
(d)   the between-group sum of squares
(e)   the within-group sum of squares

Answer: (c)
Difficulty level: Easy to Medium

10.6.1: Given the ANOVA table below, the between-group and within-group variance estimates is(are):

Source	df	SS	Variances or Mean Squares	F
Between groups	3	3000	???	??
Within groups	20	4000	???	
Total	23	7000		

(a)   3000 and 4000
(b)   1000 and 200
(c)   4000 and 3000
(d)   23 and 7000
(e)   1000/200

Answer: (b)
Difficulty level: Easy

10.6.2: Given the ANOVA table below, the F value is:

Source	df	SS	Variances or Mean Squares	F
Between groups	3	3000	???	??
Within groups	20	4000	???	
Total	23	7000		

(a)   23
(b)   7000
(c)   3000/4000
(d)   1000/200
(e)   7000/23

Answer: (d)
Difficulty level: Easy

10.6.3: Recently, many companies have attempted to increase employee job satisfaction by changing work schedules or by making work schedules more flexible. Random samples were selected from three normal populations with equal variances. The results for employee job satisfaction are presented in the table below. <u>Answer the following using</u> $\alpha = .05$.

$$\overline{X}_1 = 5, \qquad \overline{X}_2 = 2, \qquad \overline{X}_3 = 8$$

$$s_1 = 2, \qquad s_2 = 3, \qquad s_3 = 3$$

$$n_1 = 5, \qquad n_2 = 4, \qquad n_3 = 4$$

a. What is the grand mean for all of the sample data?
b. What are the between-group and within-group variance estimates?
c. What are the between-group and within-group degrees of freedom?
d. What inference should we make?

Answer: a.   Grand Mean = 5

   b. $s^2_B = 36$, $s^2_W = 7$

   c. $F = 36/7 = 5.143$, $F_{.05,2,10} = 4.1028$

   d.   Reject H_0; conclude that at least one of the population means for employee job satisfaction is different

Difficulty level: Medium

10.6.4: When the null hypothesis $H_0: \mu_1 = \mu_2 = \mu_3$ is true:

(a)   the F ratio approximately equals zero
(b)   the F ratio is likely to be near one
(c)   the F ratio is likely to be very large
(d)   the results are said to be statistically significant
(e)   the two factors are strongly correlated with one another

Answer: (b)
Difficulty level: Easy to Medium

10.6.5: Given the ANOVA table below, the between-group and within-group degrees of freedom (at α = .05) are:

Source	df	SS	Variances or Mean Squares	F
Between groups	??	300	100	??
Within groups	??	600	20	
	--	---		
Total	33	900		

(a)   100 and 20
(b)   300 and 600
(c)   3 and 30
(d)   5 and 2.9223
(e)   5 and 3.5894

Answer: (c)
Difficulty level: Easy

10.6.6: Given the ANOVA table below, the calculated and critical F values (at α = .05) are:

Source	df	SS	Variances or Mean Squares	F
Between groups	??	300	100	??
Within groups	??	600	20	
	--	---		
Total	33	900		

(a)   100 and 20
(b)   300 and 600
(c)   3 and 30
(d)   5 and 2.9223
(e)   5 and 3.5894, respectively

Answer: (d)
Difficulty level: Medium

10.6.7: Given the ANOVA table below, the appropriate decision at $\alpha = .05$ is to _____.

Source	df	SS	Variances or Mean Squares	F
Between groups	??	300	100	??
Within groups	??	600	20	
Total	33	900		

(a) fail to reject the null hypothesis and conclude that the three population means are equal
(b) fail to reject the null hypothesis and conclude that the four population means are equal
(c) reject the null hypothesis and conclude that at least one of the three population means is different
(d) reject the null hypothesis and conclude that at least one of the four population means is different
(e) fail to reject the null hypothesis and conclude that at least one of the four population means is different

Answer: (d)
Difficulty level: Medium to Challenging

10.6.8: The larger the ratio of the between-group and within-group variance estimates, the more likely it is that:

(a) we will reject the null hypothesis
(b) we will accept the null hypothesis
(c) we will commit a Type II error
(d) the population variances are equal to one another
(e) the test will be invalid due to sampling error

Answer: (a)
Difficulty level: Medium

10.6.9: An individual investor wants to know if there are
significant differences in portfolio management with respect
to the average maturity (in days) for three types of tax-
exempt money market funds. Random samples are taken from
normal populations with equal variances, with the results
presented below. The inference we should make:

	Tax-free income fund	Tax-exempt fund	Municipal Bond fund
n_j	4	4	3
$\overline{X}_j$	6.00	10.00	8.00
s_j	.500	2.00	1.50

(a) is to conclude that the mean maturity is equal for each
     of the three money market funds at α = .05
(b) is to conclude that the mean maturity is different for
     each of the three money market funds at α = .05
(c) is to conclude that the mean maturity is different for
     each of the three money market funds at α = .01
(d) is to conclude that at least one of the mean maturities
     is different at α = .01
(e) none of these

Answer: (e)
Difficulty level: Medium to Challenging

10.6.10: A retailer samples three groups of customers to ascertain preferences, if any, for particular shopping data. The data are compiled and summarized in the table below. <u>Answer the following using $\alpha = .05$.</u>

	Group 1	Group 2	Group 3
	8	12	11
	6	15	8
	10	9	11
Sum	24	36	30
Mean	8	12	10

a. What is the grand mean for all of the sample data?
b. What are the between-group and within-group variance estimates?
c. What are the calculated and critical F values?
d. What inference should we make?

Answer: a. Grand Mean = 10

b. $s^2_B = 12$, $s^2_W = 5.33$

c. $F = 12/5.33 = 2.25$, $F_{.05,2,6} = 5.1433$

d. We do not have strong enough evidence to reject the null hypothesis that the population means are all equal

Difficulty level: Easy to Medium

10.6.11: A public opinion analyst randomly samples four
groups of residents to determine whether the average number
of hours people spend watching television daily is the same
for different suburbs in the New York City area. The data
are summarized in the table below. <u>Use α = .05 to answer
the following questions</u>.

	Group 1	Group 2	Group 3	Group 4
n_j	4	4	6	6
$\overline{X}_j$	4	4	8	8
s_j	1	1	4	2

   a.   What is the grand mean for all of the sample data?
   b.   What are the between-group and within-group variance
       estimates?
   c.   What are the calculated and critical F values?
   d.   What inference should we make?
   e.   Are any of the assumptions required likely to be
       violated by the above sample data? If yes, which one?

Answer: a.   Grand Mean = 6.4

           b. $s^2_B = 25.6$, $s^2_W = 6.625$

           c. $F = 3.864$, $F_{.05,3,16} = 3.2389$

           d.   We should reject the null hypothesis and
                 conclude that at least one of the population
                 means is different
           e.   The sample variance of group 3 is much higher
                 than the others, indicating that group 3 may
                 have a higher population variance as well
Difficulty level: Medium

10.6.12: Sierra Pacific Electric Company uses water-cooled turbines at one of its power-generating plants. If the water used in the cooling system is too polluted, then the system will become corroded. For this reason, filters are used to reduce the pollution before the water enters the system. Managers at Sierra Pacific would like to test the effectiveness of four different types of filters in reducing water pollution by testing the equality of mean pollution counts when the system is used with each of the four types of filters. Random samples are taken, with the results summarized in the table below. Assume that populations are normally distributed with equal variances. <u>Answer the following using</u> $\alpha = .01$.

	Filter 1	Filter 2	Filter 3	Filter 4
n_j	3	3	3	3
$\overline{X}_j$	8	12	10	22
s_j^2	7	13	7	13

a. What is the appropriate procedure for testing the equality of mean pollution counts for the 4 filters?
b. What is the grand mean for all of the sample data?
c. What are the between-group and within-group variance estimates?
d. What are the between-group and within-group degrees of freedom?
e. What are the calculated and critical F values?
f. What inference should we make?

Answer: a. Analysis of variance.

b. Grand Mean = 13.0.

c. $s_B^2 = 116$, $s_W^2 = 10$

d. $v_1 = (4 - 1) = 3$, $v_2 = (12 - 4) = 8$

e. $F = 116/10 = 11.6$, $F_{.01,3,8} = 7.5910$

f. At $\alpha = .01$, we should reject H_0 and conclude that the population mean pollution count for at least one of the filters; namely filter 4, is different (from the other filters)

Difficulty level: Easy to Medium

10.6.13: A career consultant wants to know whether students with degrees in business administration who work in the functional areas of (1) management, (2) accounting, and (3) finance have the same interest in business details. Random samples of interest in business detail, which is measured by an index, were taken from individuals who work in the three functional areas, with the results presented in the table below. Assume that the populations are normally distributed with equal variances. <u>Answer the following using $\alpha = .05$</u>.

	Management	Accounting	Finance
n_j	4	4	3
$\overline{X}_j$	5.25	10.00	8.33
s_j	.577	2.16	1.53

a. What are the between-group and within-group variance estimates?
b. What are the calculated and critical F values?
c. What inference should we make?

Answer: a. $s^2_B = 40.643$, $s^2_W = 2.460$

b. $F = 16.52$, $F_{.05,3,8} = 4.0662$

c. We reject H_0 and conclude that students in the different fields have different degrees of interest in business detail

Difficulty level: Easy to Medium

10.9.1: Which of the following statements is true about a randomized complete block design?

(a) Items in an experiment can be separated into n homogeneous groups.
(b) Each item in a block is assigned randomly to one of the K levels of a treatment factor.
(c) Blocking often improves the precision of the experiment by decreasing the error variance without increasing the size of the experiment.
(d) Blocking is effective if the values of the dependent variable have more variation between blocks than they have within the blocks.
(e) All of these.

Answer: (e)
Difficulty level: Easy to Medium

Tests Using Categorical Data

11.1.1: The general procedure or test technique for testing the equality of several proportions is:

(a) a χ^2 test of independence
(b) a χ^2 goodness-of-fit test
(c) simple linear regression analysis
(d) a multinomial test
(e) an F test

Answer: (d)
Difficulty level: Easy

11.2.1: A survey of 600 families during a weekend prime time period was taken, with the results for numbers of viewers tuned to the various networks summarized below. If we want to test the hypothesis that each of the three major networks has 30% of the weekend prime time market ($p_1 = p_2 = p_3$) and Fox has a 10% share ($p_4 = .10$), we should (at $\alpha = .01$):

NBC	190
CBS	170
ABC	160
FOX	80

(a) Reject H_0, since the test statistic is less than 10.0
(b) Reject H_0, since the test statistic exceeds 11.34
(c) Do not reject H_0, since the test statistic is less than 11.34
(d) Do not reject H_0, since the test statistic exceeds 11.34
(e) Do not reject H_0, since the test statistic is less than 12.84

Answer: (c)
Difficulty level: Medium

11.3.1: Which of the following tests does not use the χ^2 distribution?

(a) homogeneity tests
(b) tests of independence
(c) contingency table tests
(d) test for the equality of two population variances
(e) tests for whether a population follows a normal distribution with a mean of μ

Answer: (d)
Difficulty level: Easy

11.3.2: The hypothesis tests for both the χ^2 and F
distributions are most often:

(a)   used to standardize values of the sample mean or the
      sample proportion to find areas under the curve
(b)   done as a one-sided, upper tail hypothesis test
(c)   done as a one-sided, lower tail hypothesis test
(d)   done as a two-sided hypothesis test
(e)   symmetric

Answer: (b)
Difficulty level: Easy to Medium

11.3.3: When the χ^2 test statistic is very large:

(a)   the null hypothesis must be accepted
(b)   the null hypothesis is very likely to be rejected
(c)   type II errors are large and nonrandom
(d)   sampling errors are large and nonrandom
(e)   the power of the test is very small

Answer: (b)
Difficulty level: Easy to Medium

11.3.4: The test statistic for a contingency table test will
follow a χ^2 distribution with degrees of freedom given by
the formula _____, provided that _____.

(a) $(r - 1)(c - 1)$; $np(1 - p) \geq 5$ for each comparison
(b) $(k - h - 1)$; $np(1 - p) \geq 5$ for each comparison
(c) $(r - 1)(c - 1)$; $np(1 - p) < 5$ for each comparison
(d) $(k - h - 1)$; $np(1 - p) < 5$ for each comparison
(e) $(r - 1)(c - 1)$; samples are random and independent

Answer: (a)
Difficulty level: Medium

11.3.5: Under the assumption of independence, the expected
frequencies for a χ^2 test:

(a)   must equal the observed frequencies
(b)   must exceed the observed frequencies
(c)   are computed by dividing the product of the row and
      column totals by the grand total
(d)   are computed by pooling the sample proportions
(e)   will be the same for each of the comparisons being made

Answer: (c)
Difficulty level: Easy

11.4.1: The sample size for a contingency table should be large enough so that:

(a)   expected frequencies are each at least equal to 5
(b)   observed frequencies are each at least equal to 5
(c)   expected frequencies are each at least equal to 30
(d)   observed frequencies are each at least equal to 30
(e)   the sum of the sample sizes will exceed 30

Answer: (a)
Difficulty level: Easy

11.4.2:   For each of the following χ^2 testing situations, find the degrees of freedom (at $\alpha = .05$):

a.   a test of independence with $r = 5$, $c = 4$
b.   a test of independence with $r = 3$, $c = 4$
c.   a goodness-of-fit test with $k = 15$, $h = 1$
d.   a goodness-of-fit test with $k = 8$, $h = 0$
e.   a homogeneity test with $r = 2$, $c = 2$

Answer: a.   df = $(5 - 1)(4 - 1) = 12$
           b.   df = $(3 - 1)(4 - 1) = 6$
           c.   df = $15 - 1 - 1 = 13$
           d.   df = $8 - 0 - 1 = 7$
           e.   df = $(2 - 1)(2 - 1) = 1$
Difficulty level: Easy

11.4.3:   For each of the following situations, find the critical χ^2 value for the test (at $\alpha = .01$):

a.   a test of independence with $r = 5$, $c = 4$
b.   a test of independence with $r = 3$, $c = 4$
c.   a goodness-of-fit test with $k = 15$, $h = 1$
d.   a goodness-of-fit test with $k = 8$, $h = 0$
e.   a homogeneity test with $r = 2$, $c = 2$

Answer: a.   Critical value = 26.22
           b.   Critical value = 16.81
           c.   Critical value = 27.69
           d.   Critical value = 18.48
           e.   Critical value = 16.81
Difficulty level: Easy to Medium

11.4.4: A clothing and apparel company has always produced three sizes of sports shirts: small, medium, and large. The output of its shop has always been 25% small size sports shirts, 50% medium size sports shirts, and 25% large sizes sport shirts. This breakdown was determined several years age on the basis of a guess about the true proportion of small, medium, and large men in the population. Recently, the new company president has questioned these proportions. A random sample of 200 men was selected and measured for sport shirt size. It was found that 35 men were small, 90 were medium, and 75 were large. If we test the hypothesis that $p_1 = .25$, $p_2 = .50$, and $p_3 = .25$, using the above data, the calculated and critical values for the test are (at $\alpha = .01$):

(a)   18 and 10.6
(b)   10.6 and 18
(c)   18 and 9.21
(d)   9.21 and 18
(e)   18 and $\pm$ 2.575

Answer: (c)
Difficulty level: Easy to Medium

11.4.5: A clothing and apparel company has always produced three sizes of sports shirts: small, medium, and large. The output of its shop has always been 25% small size sports shirts, 50% medium size sports shirts, and 25% large sizes sport shirts. This breakdown was determined several years age on the basis of a guess about the true proportion of small, medium, and large men in the population. Recently, the new company president has questioned these proportions. A random sample of 200 men was selected and measured for sport shirt size. It was found that 35 men were small, 90 were medium, and 75 were large. If we test the hypothesis that $p_1 = .25$, $p_2 = .50$, and $p_3 = .25$, using the above data, at $\alpha = .01$, our decision is:

(a)   do not reject H_0 since the calculated value is less than 18
(b)   do not reject H_0 since the calculated value exceeds 9.21
(c)   reject H_0 since the calculated value is less than 10.6
(d)   reject H_0 since the calculated value exceeds 9.21
(e)   reject H_0 since the calculated value is less than 18

Answer: (d)
Difficulty level: Easy to Medium

11.4.6: A recruiter from a health maintenance organization needed to hire a physician in an outpatient clinic located in a rural town. The recruiter believed that physicians' specialties were *independent* of the location of the physicians practice. A random sample of physicians was selected from a reference listing, and the location classifications and the numbers practicing in a given specialty are given in the table below. At α = .05, we should _____ the recruiter's claim because _____.

Location	Physicians' Specialty		
	Family Practice	Surgery	Internist
Rural	30	4	10
Suburban	20	40	20
Urban	5	30	15

(a) accept; the results of the test are unreliable
(b) accept; the test statistic is less than 9.49
(c) accept; the test statistic is less than 11.14
(d) reject; the test statistic exceeds 11.14
(e) reject; the test statistic exceeds 9.49

Answer: (d)
Difficulty level: Medium

11.4.7: A group of students was asked if they preferred to study alone or with a friend. 110 men and 90 women were sampled. 45 of the men and 50 of the women said that they preferred studying alone. You are interested in knowing whether preferences for studying alone (or with someone else) are different for male and female students. Answer the following using α = .05.

a. What are the appropriate null and alternative hypotheses?
b. What are the calculated and critical values?
c. What inference should we make?

Answer: a. H_0: preference for studying alone and gender are independent
H_a: preference for studying alone and gender are dependent
b. χ^2 = 4.258, $\chi^2_{.05,1}$ = 3.84
c. Since we can reject the null hypothesis of independence, we have strong evidence that student preferences for studying alone are related to student gender
Difficulty level: Easy to Medium

11.4.8: If an expected frequency for any comparison is less than 5, then:

(a) you must use the finite population correction factor
(b) you must pool the sample variances
(c) you cannot use the χ^2 distribution
(d) you must combine comparisons until each of the expected frequencies is at least 5
(e) none of these

Answer: (d)
Difficulty level: Easy

11.4.9: Which parameters does the χ^2 distribution depend on?

(a) the population mean
(b) the population variance
(c) the population proportion
(d) the degrees of freedom
(e) all of these

Answer: (d)
Difficulty level: Easy

11.4.10: If there are three types of primary vocational interests for graduating medical school students and four quartiles for the students MCAT score, the number of degrees of freedom for the χ^2 distribution is:

(a) 12
(b) 10
(c) 8
(d) 6
(e) 4

Answer: (d)
Difficulty level: Easy

11.4.11: The χ^2 test is an example of a:

(a) one-sided upper tail nonparametric test
(b) one-sided lower tail nonparametric test
(c) one-sided upper tail parametric test
(d) two-sided nonparametric test
(e) two-sided parametric test

Answer: (a)
Difficulty level: Easy

11.4.12: A metropolitan police force is experimenting with the use of two types of patrols: (1) an officer and a dog, and (2) two officers. To determine if there is a relationship between the type of patrol used and the type of crimes in an area, data were gathered and summarized in the following table. At $\alpha = .05$, for a test of the null hypothesis that the type of crime is independent of the type of patrol used, we should:

	Burglary	Domestic Quarrels	Assault/ Murder/ Rape	Total
Officer And Dog	46	72	10	128
Two Officers	154	288	70	512
Total	200	360	80	640

(a) do not reject H_0 since the calculated value is less than 5.99
(b) do not reject H_0 since the calculated value exceeds 3.94
(c) reject H_0 since the calculated value is less than 5.99
(d) reject H_0 since the calculated value exceeds 3.94
(e) reject H_0 since the calculated value exceeds 5.99

Answer: (a)
Difficulty level: Medium

11.4.13: A random sample of 1000 diners at a well-known chinese restaurant were asked what their preferred appetizer and entree were from a list of six entrees and six appetizers. For a test of the null hypothesis that the preferred appetizer is independent of the preferred entree among diners at the chinese restaurant, the degrees of freedom are:

(a) 10
(b) 36
(c) 25
(d) 11
(e) unable to be determined with the information given

Answer: (c)
Difficulty level: Easy

11.4.14: A special interest group wants to determine whether a person's party affiliation and political ideology are related. A random sample of 100 people yielded the results summarized in the table below. At $\alpha = .05$, the calculated and critical values of the test are:

	POLITICAL IDEOLOGY		
PARTY AFFILIATION	Liberal	Moderate	Conservative
Democrat	15	20	5
Republican	10	10	20
Independent	5	10	5

(a) 14.58 and 9.49
(b) 14.58 and 11.14
(c) 9.49 and 14.58
(d) 11.14 and 14.58
(e) 14.58 and ± 1.96

Answer: (a)
Difficulty level: Easy to Medium

11.4.15: A special interest group wants to determine whether a person's party affiliation and political ideology are related. A random sample of 100 people yielded the results summarized in the table below. At $\alpha = .05$, we should:

	POLITICAL IDEOLOGY		
PARTY AFFILIATION	Liberal	Moderate	Conservative
Democrat	15	20	5
Republican	10	10	20
Independent	5	10	5

(a) do not reject H_0, since the calculated value is less than 14.58
(b) do not reject H_0, since the calculated value is less than 9.49
(c) reject H_0, since the calculated value exceeds 14.58
(d) reject H_0, since the calculated value exceeds 9.49
(e) do nothing, since the results are inconclusive

Answer: (d)
Difficulty level: Easy to Medium

11.4.16: A sociologist believes that intelligence might be associated with air pollution. To determine whether such a dependency exists, a random sample of 200 is taken, with the results summarized in the table below. At value of the test statistic and appropriate decision (at α = .01) are:

	IQ SCORE		
POLLUTION LEVEL	High	Medium	Low
High	35	50	15
Low	25	50	25

(a) 9.21, do not reject the null hypothesis of independence
(b) 4.17, do not reject the null hypothesis of independence
(c) 9.21, reject the null hypothesis of independence
(d) 4.17, reject the null hypothesis of independence
(e) 4.17, do not reject the null hypothesis that the population is normally distributed

Answer: (b)
Difficulty level: Easy to Medium

11.4.17: 200 fur bearing animals were damaged by a recent oil spill. Wildlife experts work to save the animals by cleaning the oil from them. A previously untried chemical is used on 100 of the animals in an attempt to save them. Given the results summarized in the table below, if we want to test whether the chemical used and animal survival are independent, our conclusion (at α = .05) should be to:

	FATE OF THE ANIMAL	
CHEMICAL USED	Animal Survives	Animal Dies
Old Chemical	60	40
New Chemical	80	20

(a) do not reject H_0 and conclude that chemical used and animal survival are independent
(b) do not reject H_0 and conclude that chemical used and animal survival are dependent
(c) reject H_0 and conclude that the population follows a Poisson distribution
(d) reject H_0 and conclude that chemical used and animal survival are dependent
(e) reject H_0 and conclude that the population follows a normal distribution

Answer: (d)
Difficulty level: Easy to Medium

11.4.18: A university located in a small town was interested in determining the relationship between residents attending university functions and affiliation with the school. A random sample of 400 residents was taken, with the results summarized in the contingency table below. <u>Answer the following using</u> $\alpha = .05$.

	Attend University Functions	Do Not Attend University Functions	Total
Affiliated	85	35	120
Not Affiliated	15	265	280
Total	100	300	400

a. What are the appropriate null and alternative hypotheses?
b. What are the calculated and critical values?
c. What inference should you make?

Answer: a. H_0: the two factors are independent
H_1: the two factors are dependent
b. $\chi^2 = 192.06$, $\chi^2_{.05,1} = 3.84$
c. Reject H_0 and conclude that there is a significant relationship between university affiliation and attending University functions
Difficulty level: Easy to Medium

11.5.1: At one time the gender ratio in an occupation was 8 women to every man. Suppose that a random sample of 450 workers contained 68 men. At $\alpha = .01$, would we be justified in concluding that the ratio changed?

(a) Yes, since the calculated value is less than 7.29
(b) Yes, since the calculated value exceeds 6.63
(c) Yes, since the calculated value is less than 6.63
(d) No, since the calculated value exceeds 6.63
(e) No, since the calculated value is less than 6.63

Answer: (b)
Difficulty level: Medium

11.5.2: A random sample of 200 truck drivers had 62 who had
been involved in at least one accident. At α = .01, if we
test H_0: p = .75, where p is the proportion of *accident-free*
drivers, the value of the χ^2 test statistic is:

(a) 3.84
(b) 1
(c) 6.63
(d) 1.96
(e) unable to be determined with the information given

Answer: (a)
Difficulty level: Medium

11.6.1: The value of the χ^2 test statistic for the null
hypothesis that the following data were drawn from a Poisson
distribution with a mean rate of 1 (at α = .01) is:

number of occurrences	frequency
0	32
1	40
2 or more	28

(a) .9983
(b) 5.9257
(c) 9.21
(d) 10.60
(e) unable to be determined with the information given

Answer: (a)
Difficulty level: Medium

11.6.2: If we test the null hypothesis that the following
data were drawn from a Poisson distribution with a mean rate
of 1 at α = .05, we should conclude that:

number of occurrences	frequency
0	32
1	40
2 or more	28

(a) likely follows a Poisson distribution
(b) likely does not follow a Poisson distribution
(c) likely follows a normal distribution
(d) likely does not follow a normal distribution
(e) is based on a discrete random variable

Answer: (a)
Difficulty level: Medium

11.6.3: The value of the χ^2 test statistic for the null
hypothesis that the following data were drawn from a normal
population with a $\mu = 20$ and $\sigma = 2$ (at $\alpha = .10$) is:

number of occurrences	frequency
less than 18	14
18 less than 20	35
20 less than 22	41
more than 22	10

(a) 3.7966
(b) 6.25
(c) 7.81
(d) 2.1232
(e) unable to be determined with the information given

Answer: (d)
Difficulty level: Medium

11.6.4: A study on airport safety was conducted. A total of
150 airports were sampled, yielding the results presented
below. If we want to test whether the number of annual
hijackings follows a Poisson distribution with a mean rate
of 2.4 hijackings/year, we should (at $\alpha = .05$):

Number of Annual Hijackings	Number of Airports
0	17
1	34
2	45
3	23
4	10
5	9
6 or more	12

(a) Reject H_0, since the calculated value exceeds 16.09
(b) Reject H_0, since the calculated value exceeds 11.069
(c) Do not reject H_0, since the calculated value is less
     than 16.09
(d) Do not reject H_0, since the calculated value is less
     than 11.069
(e) Do nothing, since the test results are inconclusive

Answer: (b)
Difficulty level: Medium

11.6.5: You are interested in determining whether the true average income of residents in a region of the eastern United States follows a normal distribution with a mean of μ = $30,000 and a standard deviation of σ = $10,000. A random sample of 500 residents is taken, yielding the results summarized in the frequency distribution below. <u>Answer the following questions using α = .05</u>.

ANNUAL INCOME RANGE	OBSERVED FREQUENCY
Less than $5,000	5
$ 5000 less than $10,000	10
$10,000 less than $20,000	70
$20,000 less than $30,000	175
$30,000 less than $40,000	175
$40,000 less than $50,000	55
$50,000 less than $60,000	5
At least $60,000	5

a.   What are the appropriate null and alternative hypotheses?
b.   What are degrees of freedom for the above test?
c.   What are the calculated and critical values?
d.   What inference should we make?
e.   Comment on how confident you are in making your decision about the distribution of plane arrivals, given the above data and procedures for doing the test.

Answer: a. H_0: the population is normal with parameters μ = $30,000 and σ = $10,000
                 H_a: the population is not normal with parameters μ = $30,000 and σ = $10,000

           b.   Degrees of freedom = 5, since we only have 6 comparisons after we collapse classes using the rule that each e_i must be at least 5

           c. χ^2 = 4.06, $\chi^2_{.05,5}$ = 11.14

           d.   We cannot reject H_0. We therefore do not have strong evidence against the hypothesis that the population is normally distributed with parameters μ = $30,000 and σ = $10,000

           e.   We should be cautious before concluding that the distribution of annual incomes *must* be normally distributed with a mean of $30,000 and a standard deviation of $10,000. While there is (weak) evidence in support of the null hypothesis, we should note that failure to combine classes results in rejecting H_0.

Difficulty level: Easy to Challenging

11.6.6: An air traffic control analyst is interested in
knowing whether the number of plane arrivals during 4:00
p.m. to 6:00 p.m. follows a Poisson distribution with a mean
rate of 1.5 plane arrivals/minute. Based on a random sample
of 48 time intervals, each of which lasting *five* minutes,
the following results, summarized in the table below, are
obtained. Answer the following questions using $\alpha = .05$.

NUMBER OF PLANE ARRIVALS	OBSERVED FREQUENCY
0 Less than 3	3
3 Less than 6	9
6 Less than 9	19
9 Less than 12	11
12 Less than 15	3
15 Less than 18	3

a.   What are the appropriate null and alternative
     hypotheses?
b.   What are degrees of freedom for the above test?
c.   What are the calculated and critical values?
d.   What inference should we make?
e.   Comment on how confident you are in making your
     decision about the distribution of plane arrivals,
     given the above data and procedures for doing the test.

Answer: a. H_0:   the population follows a Poisson
                         distribution with a mean rate of 7.5
                         plane arrivals every five minutes
                H_a: the population does not follow a Poisson
                         distribution with a mean rate of 7.5
                         plane arrivals every five minutes

           b.   Degrees of freedom = 2, since we only have 3
                comparisons after we collapse classes using
                the rule that each e_i must be at least 5

           c. $\chi^2 = .121$, $\chi^2_{.05,2} = 5.99$

           d.   We cannot reject the null hypothesis. We do
                not have string evidence against the
                hypothesis that the population follows a
                Poisson distribution with a mean rate of 7.5
                planes every five minutes

           e.   We should be cautious before concluding that
                the population *must* follow the Poisson
                distribution (and not a different one).
                While there is (weak) evidence in support of
                the null hypothesis, we would have rejected
                H_0 if we didn't combine classes with e_i's < 5.
Difficulty level: Easy to Challenging

Regression and Correlation

12.1.1: In regression analysis, the variable whose value is always known and which is often used to estimate the value of a related variable whose value is not known, is the:

(a)   dummy variable
(b)   linear variable
(c)   dependent variable
(d)   independent variable
(e)   multiple variable

Answer: (d)
Difficulty level: Easy

12.1.2: Regression analysis is the appropriate statistical technique for each of the following situations *except*:

(a)   developing a prediction equation to better predict annual sales levels for given advertising expenditure levels
(b)   studying the relationship between an individual's educational level and their annual income
(c)   forecasting future interest rates from an equation relating interest rates and changes in bond prices
(d)   all of these situations are designed specifically for regression analysis
(e)   estimating a functional relationship between the weight and gas mileage of compact automobiles

Answer: (b)
Difficulty level: Easy to Medium

12.1.3: The main objective of regression analysis is:

(a)   to accurately predict values of a dependent variable given values of an independent variable
(b)   to analyze the strength and direction of the linear relationship between a dependent variable and an independent variable
(c)   to assess goodness of fit
(d)   to obtain a statistically significant linear relationship between a dependent variable and an independent variable
(e)   to obtain a strong causal relationship between a dependent variable and an independent variable

Answer: (a)
Difficulty level: Easy to Medium

12.1.4: In the linear regression model: $Y_i = ß_0 + ß_1X_i + \epsilon_i$, our primary objective is:

(a)   to accurately predict values of Y given values of X
(b)   to analyze the strength and direction of the linear relationship between Y and X
(c)   to minimize the sum of squared errors (SSE)
(d)   to obtain a statistically significant linear relationship between Y and X
(e)   to make the necessary data transformations so that a strong causal relationship between a dependent variable and an independent variable is obtained

Answer: (a)
Difficulty level: Easy

12.1.5: In the linear regression model: $Y_i = ß_0 + ß_1X_i + \epsilon_i$, the terms $ß_0$ and $ß_1$ represent:

(a)   the Y intercept and the slope of the population regression line
(b)   the Y intercept and the slope of the sample regression line
(c)   statistics used to estimate the parameters b_0 and b_1
(d)   parameters used to estimate the parameters b_0 and b_1
(e)   the value for the random fluctuation or error

Answer: (a)
Difficulty level: Easy

12.1.6: In the linear regression model: $Y_i = ß_0 + ß_1X_i + \epsilon_i$, the assumptions necessary to obtain best linear unbiased estimates for the parameters $ß_0$ and $ß_1$ include:

I.   X values are known and not random.
II.   For each value of X, Y is normally and independently distributed with mean $\mu_{Y|X}$ equal to $(ß_0 + ß_1X)$ and variance $\sigma^2_{Y|X}$.
III.   For each X the variance of Y is the same; hence, there is no heteroscedasticity.

(a)   Only assumption I is needed
(b)   Only assumption II is needed
(c)   Only assumption III is needed
(d)   Only assumptions II and III are needed
(e)   Assumptions I, II, and III are all needed

Answer: (e)
Difficulty level: Easy

12.1.7: In the linear regression model: $Y_i = \beta_0 + \beta_1 X_i + \epsilon_i$, the term ϵ represents:

(a) the Y intercept of the population regression line
(b) the slope of the population regression line
(c) a statistic used to estimate the parameter e
(d) a parameter used to estimate the statistic e
(e) the value for the random fluctuation or error

Answer: (e)
Difficulty level: Easy

12.1.8: A(n) _____ is a convenient means often used to help determine the appropriate functional form of the relationship between two variables.

(a) scatterplot
(b) Durbin-Watson test
(c) Venn diagram
(d) F test
(e) correlation matrix

Answer: (a)
Difficulty level: Easy

12.2.1: In linear regression, the estimates for b_1 and b_2 that are found using the method of least squares:

(a) are necessarily equal to β_0 and β_1
(b) are interval estimates for β_0 and β_1
(c) are best linear unbiased estimates for β_0 and β_1, regardless of whether any of the assumptions are violated
(d) take into account nonlinear patterns in the data
(e) none of these

Answer: (e)
Difficulty level: Easy to Medium

12.2.2: The method used by statisticians to find the best linear fit through a set of data points is called:

(a)   a scatterplot
(b)   regression analysis
(c)   correlation analysis
(d)   the method of least squares
(e)   the coefficient of determination

Answer: (d)
Difficulty level: Easy

12.2.3: The strength and direction of the linear relationship between a person's educational level and their annual income is best analyzed using:

(a)   simple linear regression analysis
(b)   multiple regression analysis
(c)   correlation analysis
(d)   a correlation matrix
(e)   a scatterplot

Answer: (c)
Difficulty level: Easy to Medium

12.2.4: Which of the following criteria are satisfied by the method of least squares?

I.   the sum of the squared deviations between the observed and predicted values of the dependent variable (SSE) is minimized
II.   the sum of the deviations between the observed and predicted values of the dependent variable equals zero
III. the coefficient of determination r^2 is minimized

(a)   I only
(b)   II only
(c)   I and III only
(d)   I and II only
(e)   I, II and III all must be satisfied

Answer: (d)
Difficulty level: Easy to Medium

12.2.5: Using the method of least squares to find the best linear fit through a set of data points results in values for the Y intercept and the slope which minimize the sum of squared deviations between:

(a)   the vertical differences of the actual and expected values of Y
(b)   the horizontal differences of the actual and expected values of Y
(c)   the vertical differences of the actual and predicted values of Y
(d)   the horizontal differences of the actual and predicted values of Y
(e)   the values of X and Y

Answer: (c)
Difficulty level: Easy to Medium

12.2.6: Given the data values for X and Y below, the least squares estimates for the Y-intercept and the slope are:

Y	36	80	44	55	35
X	9	15	10	11	10

(a)   50 and 11
(b)   169 and 22
(c)   7.68 and -34.5
(d)   7.68 and 39
(e)   unable to be determined with the information given

Answer: (c)
Difficulty level: Easy to Medium

12.2.7: Given the data values for X and Y below, the predicted value for Y, given that X = 10, is:

Y	36	80	44	55	35
X	9	15	10	11	10

(a)   42.3
(b)   76.8
(c)   111.3
(d)   -34.5
(e)   unable to be determined with the information given

Answer: (a)
Difficulty level: Medium

12.2.8: Given the following data values for annual sales (Y) and unit price (X), where Y is measured in thousands of dollars and X is measured in hundreds of dollars, the least squares estimate for the slope coefficient is:

Y	45	55	62	90	73
X	32	28	25	14	21

(a) -2.51
(b) $60,000
(c) $2400
(d) 71.0253
(e) $71,025.3

Answer: (a)
Difficulty level: Easy to Medium

12.2.9: Given the following data values for annual sales (Y) and unit price (X), where Y is measured in thousands of dollars and X is measured in hundreds of dollars, the predicted value for Y, given a unit price of $2600, is:

Y	45	55	62	90	73
X	32	28	25	14	21

(a) about $60,000
(b) about $71,000
(c) about $6,000
(d) about $71
(e) about $60

Answer: (a)
Difficulty level: Medium

12.3.1: In a regression analysis if SSR = 100 and SSE = 25, then the sample coefficient of determination equals:

(a) .80
(b) .8944
(c) .20
(d) .4472
(e) .64

Answer: (a)
Difficulty level: Easy

12.3.2: In a regression analysis if SSR = 100 and SSE = 25, then the sample correlation coefficient:

(a)   .80
(b)   .8944
(c)   .20
(d)   .4472
(e)   .64

Answer: (b)
Difficulty level: Easy

12.3.3: If the sample coefficient of determination has a positive value, then we know that:

(a)   the sample t value must be positive
(b)   the population correlation coefficient may be positive or negative, depending on the sample information
(c)   the sample correlation coefficient may be positive or negative, depending on the sample information
(d)   the sample slope coefficient must be positive
(e)   the sample Y intercept must be positive

Answer: (c)
Difficulty level: Easy to Medium

12.3.4: In regression analysis, the total sum of squares may be partitioned into:

(a)   SST + SSE
(b)   SSR - SSE
(c)   SST + SSR
(d)   SSE + SSR
(e)   None of these

Answer: (d)
Difficulty level: Easy

12.3.5: Which of the following are additive?

I.   Degrees of freedom
II. Mean squares
III. Sum of squares

(a) only I is additive
(b) only II is additive
(c) only III is additive
(d) only I and III are additive
(e) I, II, and III are all additive

Answer: (d)
Difficulty level: Easy to Medium

12.3.6: If we square the portion of the total deviation that is <u>unexplained</u> across all observations, the resulting measure:

(a) is called the sample coefficient of determination
(b) is called the sample correlation coefficient
(c) is the regression sum of squares (SSR)
(d) is the error sum of squares (SSE)
(e) is the total sum of squares (SST)

Answer: (d)
Difficulty level: Easy to Medium

12.3.7: The relation between MSR and SSR in a simple linear regression model:

(a) MSR equals SSR
(b) MSR equals SSR divided by 2
(c) MSR equals SSR divided by (k - 1)
(d) MSR equals SSR divided by (n - 1)
(e) MSR equals SSR divided by (n - 2)

Answer: (a)
Difficulty level: Medium

12.3.8: The relation between MSE and SSE in a simple linear regression model:

(a)   MSE equals SSE
(b)   MSE equals SSE divided by 2
(c)   MSE equals SSE divided by (k - 1)
(d)   MSE equals SSE divided by (n - 1)
(e)   MSE equals SSE divided by (n - 2)

Answer: (e)
Difficulty level: Medium

12.3.9: The relation between SSE and r^2 in a simple linear regression model:

(a)   is always a positive and significant one
(b)   is such that their sum must equal SST
(c)   is such that their sum must equal one
(d)   is such that $(1 - r^2) = [SSE/SSR]$
(e)   is such that $r^2 = 1 - [SSE/SST]$

Answer: (e)
Difficulty level: Medium

12.4.1: If the sample correlation coefficient is <u>negative</u>, then we know that:

(a)   the sample coefficient of determination must be negative
(b)   the population correlation coefficient must be negative
(c)   the sample slope coefficient must be negative
(d)   the sample slope coefficient may be positive or negative
(e)   the sample t statistic may be positive or negative

Answer: (c)
Difficulty level: Medium

12.4.2: If the sample correlation coefficient is -.70, then the percentage of variability in the dependent variable that is accounted for by the regression equation:

(a)   is 51%
(b)   is 49%
(c)   is approximately 83.61%
(d)   is approximately 16.39%
(e)   cannot be determined

Answer: (b)
Difficulty level: Easy

12.4.3: In regression and correlation analysis, if SST and SSE are both known, then with this information alone:

(a)   the sample Y intercept can be computed
(b)   the sample coefficient of determination can be computed
(c)   the sample slope coefficient can be computed
(d)   the sample t value for testing H_0: $ß_1$ = 0 can be computed
(e)   we can compute all of these if we know SST and SSE

Answer: (b)
Difficulty level: Easy

12.4.4: The sum of squared errors (SSE) that is minimized by using the method of least squares can never exceed:

(a)   the regression sum of squares (SSR)
(b)   the total sum of squares (SST)
(c)   the sample coefficient of determination (r^2)
(d)   the sample correlation coefficient (r)
(e)   the sample slope coefficient (b_1)

Answer: (b)
Difficulty level: Easy

12.4.5: If the sample coefficient of determination is <u>negative</u>, then we know that:

(a)   the population correlation coefficient must be negative
(b)   the sample correlation coefficient must be negative
(c)   the sample slope coefficient must be negative
(d)   the sample t statistic must be negative
(e)   the sample coefficient of determination cannot be negative

Answer: (e)
Difficulty level: Medium

12.4.6: If the coefficient of determination is positive, then the slope of the sample regression line:

(a)   must be positive
(b)   must be negative
(c)   will be a number that ranges from 0 to 1
(d)   will be positive or negative, depending on whether the sample Y intercept is positive or negative
(e)   will be positive or negative, depending on whether the sample correlation coefficient is positive or negative

Answer: (e)
Difficulty level: Easy to Medium

12.4.7: Given the analysis of variance table below, the sample coefficient of determination is:

Source of Variation	Degrees of Freedom	Sum of Squares	Mean Square	F
Regression	1	50	50	10
Error	6	30	5	
Total	7	80		

(a)   50
(b)   .625
(c)   .7906
(d)   .375
(e)   7.07

Answer: (b)
Difficulty level: Easy to Medium

12.4.8: Given the analysis of variance table below, the standard error of the estimate $s_{y|x}$ is:

Source of Variation	Degrees of Freedom	Sum of Squares	Mean Square	F
Regression	1	50	50	10
Error	6	30	5	
Total	7	80		

(a)   5
(b)   .625
(c)   .7906
(d)   .375
(e)   2.236

Answer: (e)
Difficulty level: Medium

12.4.9: Given the analysis of variance table below, the sample coefficient of determination is:

Source of Variation	Degrees of Freedom	Sum of Squares	Mean Square	F
Regression	1	???	120	5.00
Error	??	???	24	
Total	11	???		

(a)   120
(b)   10.9545
(c)   .8333
(d)   .3333
(e)   unable to be determined with the information given

Answer: (d)
Difficulty level: Medium to Challenging

12.4.10: Given the analysis of variance table below, the standard error of the estimate $s_{Y|x}$ is:

Source of Variation	Degrees of Freedom	Sum of Squares	Mean Square	F
Regression	1	???	120	5.00
Error	??	???	24	
Total	11	???		

(a) 24
(b) 4.899
(c) .8333
(d) .3333
(e) unable to be determined with the information given

Answer: (b)
Difficulty level: Medium to Challenging

12.4.11: Given the analysis of variance table below, the absolute value of the t statistic for testing H_0: $\rho = 0$ is:

Source of Variation	Degrees of Freedom	Sum of Squares	Mean Square	F
Regression	1	???	120	5.00
Error	??	???	24	
Total	11	360		

(a) 5.00
(b) 2.236
(c) ± 2.228
(d) 1.812
(e) unable to be determined with the information given

Answer: (b)
Difficulty level: Medium to Challenging

12.4.12: Given the analysis of variance table below, if testing H_0: $\rho = 0$, the appropriate decision at $\alpha = .05$ is:

Source of Variation	Degrees of Freedom	Sum of Squares	Mean Square	F
Regression	1	???	120	5.00
Error	??	???	24	
Total	11	360		

(a) do not reject H_0 since the test statistic is less than 2.228
(b) do not reject H_0 since the test statistic exceeds 1.812
(c) reject H_0 since the test statistic is less than 5.00
(d) reject H_0 since the test statistic exceeds 2.228
(e) unable to be determined with the information given

Answer: (d)
Difficulty level: Medium to Challenging

Questions 12.5.1 - 12.5.3 refer to a simple regression whose results are summarized in the computer output below.

The manager for a large chain of Italian food restaurants believes that the annual sales revenues (Y, measured in thousands of dollars) is related to the size of the student population on a nearby campus (X, measured in thousands of students). A sample of 10 restaurants is taken, yielding:

Source of Variation	Degrees of Freedom	Sum of Squares	Mean Square	F
Regression	1	14,200	14,200	74.25
Error	8	1530	191.25	
Total	9			

Variable	coefficient	s.e. of coeff.	t-ratio
Constant	60.00	9.226	6.50
X	5.00	0.5803	8.62

12.5.1: The predicted value of annual sales for a student population of 3500 students (in dollars) is:

(a) $77.50
(b) $17.50
(c) $77,500
(d) $17,500
(e) $17,560,000

Answer: (c)
Difficulty level: Easy

12.5.2: The value of the standard error of the estimate is:

(a) 9.226
(b) 0.5803
(c) 191.25
(d) 13.829
(e) unable to be determined with the information given

Answer: (d)
Difficulty level: Medium

12.5.3: At $\alpha = .05$, we should conclude that:

(a) the independent variable is statistically significant
(b) the assumption of linearity is not a good one
(c) higher student populations will lower sales revenues
(d) student population has no affect on sales revenue
(e) the error terms are highly correlated

Answer: (a)
Difficulty level: Medium

12.5.4: Given the regression equation,

Yhat = -2.45 + 0.32X, r^2 = .76, $s_{Y|X}$ = 1.03, n = 26
        (1.85)   (0.11)
with standard errors reported in parentheses, we may
conclude that:

(a) the independent variable is insignificant at $\alpha = .10$
(b) the independent variable is significant at $\alpha = .01$
(c) the average prediction error is very large
(d) there is a weak inverse linear relation between X and Y
(e) none of these

Answer: (b)
Difficulty level: Medium

12.5.5: Given the regression equation,

Yhat = -2.45 + 0.32X, r^2 = .76, $s_{Y|X}$ = 1.03, n = 26
       (1.85) (0.11)
with standard errors reported in parentheses, the predicted value of Y, given X_0 = 12, is:

(a)   -2.45
(b)   .32
(c)   3.84
(d)   1.39
(e)   none of these

Answer: (d)
Difficulty level: Easy

12.5.6: Given the regression equation,

Yhat = 5.078 - 0.76X, r^2 = .20, $s_{Y|X}$ = 4.58, n = 18
      (3.85) (1.14)
with standard errors reported in parentheses, we may conclude (at α = .01):

(a)   that the independent variable is significant
(b)   that the linear relationship between X and Y is strong
(c)   that the regression model does well at predicting
      values of Y given values of X
(d)   that the sample correlation coefficient is negative
(e)   none of these

Answer: (d)
Difficulty level: Medium

12.5.7: Find the lower and upper confidence limits of a 95% prediction interval for a *new* individual value Y_0 given that X_0 = 10 when the estimated regression line is Yhat = 16.4 - .32X, the sample size is 18, the standard error of the estimate is 4, the mean of X is 14, and Σx^2 = 3816.

(a)   8.984 to 17.416
(b)   5.839 to 20.561
(c)   4.262 to 22.138
(d)   4.937 to 21.463
(e)   12.641 to 13.759

Answer: (c)
Difficulty level: Medium

12.5.8: Find the lower and upper confidence limits of a 95% confidence interval for the *mean* value of Y given that X_0 = 16 when the estimated regression line is Yhat = 40 - .62X, the sample size is 10, the standard error of the estimate is 5, the mean of X is 13, and Σx^2 = 2000.

(a)   26.560 to 33.600
(b)   25.938 to 34.222
(c)   29.252 to 30.908
(d)   29.412 to 30.748
(e)   26.739 to 33.421

Answer: (b)
Difficulty level: Medium

12.5.9: Find the lower and upper confidence limits of a 99% confidence interval for the *mean* value of Y given that X_0 = 20 when the estimated regression line is Yhat = 40 - .62X, the sample size is 15, the standard error of the estimate is 6, the mean of X is 10, and Σx^2 = 3000.

(a)   21.001 to 34.199
(b)   25.409 to 29.791
(c)   21.958 to 33.242
(d)   22.495 to 32.705
(e)   21.794 to 33.406

Answer: (a)
Difficulty level: Medium

12.6.1: Of the following sample correlation coefficients, the one indicative of the lowest extent of linear relationship is:

(a)   -.85
(b)   -.05
(c)   +.10
(d)   +.50
(e)   +.90

Answer: (b)
Difficulty level: Easy to Medium

12.6.2: When we say that a sample correlation coefficient is significantly different from 0, we are suggesting that:

(a) it has practical value
(b) the population correlation coefficient is possibly not 0
(c) it has a positive rather than a negative sign
(d) all of the above
(e) none of the above

Answer: (b)
Difficulty level: Easy to Medium

12.6.3: Which of the following statements about ρ is false?

(a)   It measures only linear relationships.
(b)   It is the population correlation coefficient.
(c)   The variables may be perfectly related to one another but ρ can still be zero.
(d)   It ranges from 1 to -1.
(e)   None of these statements are false.

Answer: (e)
Difficulty level: Medium

12.6.4: An economist is *primarily* interested in studying the relationship between the way families spend tax refunds and the size of their annual incomes. He took a sample of six families and obtained data values for (1) the percentage of the tax refund spent within three months of receipt, and (2) the annual family income. Which of the following would be most useful for the economist to know?

(a)   the value of the sample coefficient of determination
(b)   whether the variable is significant
(c)   whether any of the violations of the simple linear regression model are violated
(d)   the value of the sample correlation coefficient
(e)   whether there is a positive or negative relationship between the two variables

Answer: (d)
Difficulty level: Medium

12.6.5: An economist is interested in studying the relation between the way families spend tax refunds and the size of their annual incomes. A sample of six families was taken, with the results summarized in the table below. What is the value of the sample correlation coefficient?

Percent of Refund, Y (expressed as a percent)	Annual Family Income, X (in thousands of dollars)
70.0	13.0
55.0	18.0
100.0	9.0
40.0	25.0
15.0	36.0
20.0	19.0

(a)   -1280
(b)   456
(c)   -.9138
(d)   -.835
(e)   .6972

Answer: (d)
Difficulty level: Easy to Medium

12.6.6: If you are interested in testing the null hypothesis that there is no relationship between variable X and variable Y, which of the following would be appropriate for doing this?

I.   testing the hypothesis that $\beta_1 = 0$, using a t-test
II.   testing the hypothesis that $\rho = 0$, using a t-test
III. testing the hypothesis that the population follows a
      normal distribution with parameters μ and σ

(a)   II only
(b)   III only
(c)   I and II only
(d)   I and III only
(e)   I, II, and III

Answer: (c)
Difficulty level: Easy to Medium

12.6.7: Given the analysis of variance table below, the value of the sample correlation coefficient:

Source of Variation	Degrees of Freedom	Sum of Squares	Mean Square	F
Regression	1	50	50	10
Error	6	30	5	
Total	7	80		

(a) is 7.07
(b) is .625
(c) must be +.7906
(d) must be -.7906
(e) is either +.7906 or -.7906

Answer: (e)
Difficulty level: Medium

12.6.8: Given the analysis of variance table below, the value of the test statistic for the test H_0: $\rho = 0$ (at $\alpha = .01$):

Source of Variation	Degrees of Freedom	Sum of Squares	Mean Square	F
Regression	1	20	20	4.00
Error	16	80	5	
Total	17	100		

(a) must be -2.00
(b) must be +2.00
(c) may be either +2.00 or -2.00
(d) is 2.583
(e) is 2.921

Answer: (c)
Difficulty level: Medium

12.6.9: A negative correlation means that:

(a) as one variable increases, the other likewise increases
(b) as one variable decreases, the other likewise decreases
(c) as one variable increases, the other variable decreases
(d) there is a curvilinear relationship between variables
(e) there is a causal relationship between variables

Answer: (c)
Difficulty level: Easy

12.6.10: If we want to know whether there exists a strong linear relationship between two variables, and a sample of 22 observations yields a sample correlation coefficient of -.50, we should (at $\alpha = .01$):

(a)   do not reject H_0 and conclude that there is not strong evidence that $\rho \neq 0$
(b)   do not reject H_0 and conclude that there is strong evidence that $\rho \neq 0$
(c)   reject H_0 and conclude that $\rho = 0$
(d)   reject H_0 and conclude that $\rho \neq 0$
(e)   maintain the status quo as the results are inconclusive

Answer: (a)
Difficulty level: Medium

12.6.11: Given a sample size of 18 and a sample correlation coefficient of -.60, if we are testing the null hypothesis H_0: $\rho = 0$ at $\alpha = .05$, we should:

(a)   do not reject H_0 and conclude that $\rho = 0$
(b)   do not reject H_0 and conclude that $\rho \neq 0$
(c)   reject H_0 and conclude that $\rho = 0$
(d)   reject H_0 and conclude that $\rho \neq 0$
(e)   maintain the status quo as the results are inconclusive

Answer: (d)
Difficulty level: Medium

12.7.1: Given the assumption that X and Y are <u>positively</u>
related, use the analysis of variance table below to compute
or answer the following questions.

Source of Variation	Degrees of Freedom	Sum of Squares	Mean Square	F
Regression	1	75	75	1.50
Error	9	450	50	
Total	10	525		

a.   What are the values of SST, SSR, and SSE?
b.   What are the values of the sample coefficient of
     determination and the correlation coefficient?
c.   What are the values of MSR, MSE, and $s_{Y|X}$?
d.   What are the calculated and critical t values for
     testing whether $H_0: \rho = 0$ (at $\alpha = .05$)?
e.   Is the linear relation between X and Y strong at $\alpha = .05$?

Answer: a.   SST = 525, SSR = 75, SSE = 450
           b. r^2 = .1429, r = + .378
           c.   MSR = 75, MSE = 50, $s_{Y|X}$ = 7.07
           d.   t = 1.225, $t_{.025,9}$ = 2.262
           e.   No, since we cannot reject H_0 at $\alpha = .05$
Difficulty level: Easy to Medium

12.7.2: Given the assumption that X and Y are <u>negatively</u>
related, use the analysis of variance table below to compute
or answer the following questions.

Source of Variation	Degrees of Freedom	Sum of Squares	Mean Square	F
Regression	1	250	250	25.0
Error	19	190	10	
Total	20	440		

a.   What are the values of SST, SSR, and SSE?
b.   What are the values of the sample coefficient of
     determination and the correlation coefficient?
c.   What are the values of MSR, MSE, and $s_{Y|X}$?
d.   What are the calculated and critical t values for
     testing whether $H_0: \rho = 0$ (at $\alpha = .01$)?
e.   Is the linear relation between X and Y strong?

Answer: a.   SST = 440, SSR = 250, SSE = 190
           b. r^2 = .5682, r = - .7538
           c.   MSR = 250, MSE = 10, $s_{Y|X}$ = 3.1623
           d.   t = -5.00, $t_{.005,19}$ = 2.861
           e.   Yes, since we can reject $H_0: \rho = 0$ at $\alpha = .01$
Difficulty level: Easy to Medium

12.7.3: Given the assumption that X and Y are <u>positively</u> related, use the analysis of variance table below to compute or answer the following questions.

Source of Variation	Degrees of Freedom	Sum of Squares	Mean Square	F
Regression	??	120	???	6.00
Error	??	400	???	
Total	21	520		

a. What are the values of SST, SSR, and SSE?
b. What are the values of the sample coefficient of determination and the correlation coefficient?
c. What are the values of MSR, MSE, and $s_{Y|X}$?
d. How accurately does the above simple regression model predict values of Y given X?
e. Is the linear relation between X and Y strong at α = .05?

Answer: a. SST = 520, SSR = 120, SSE = 400
         b. r^2 = .2308, r = + .4804
         c. MSR = 120, MSE = 20, $s_{Y|X}$ = 4.4721
         d. The proportion of total variability in Y that is explained by the regression model is low (only 23%). However, the standard error of the estimate is fairly small (only 4.47), so that average prediction errors are small.
         e. Since the t statistic of 2.45 exceeds the critical value of $t_{.025,\ 20}$ = 2.086, we can conclude that there is a strong linear relation between X and Y (at α = .05)

Difficulty level: Medium

12.7.4: Given the assumption that X and Y are <u>negatively</u> related, use the analysis of variance table below to compute or answer the following questions.

Source of Variation	Degrees of Freedom	Sum of Squares	Mean Square	F
Regression	1	250	???	??
Error	??	???	25	
Total	25	???		

a. What are the values of SST, SSR, and SSE?
b. What are the values of the sample coefficient of determination and the correlation coefficient?
c. What are the values of MSR, MSE, and $s_{Y|X}$?

Answer: a. SST = 850, SSR = 250, SSE = 600
         b. r^2 = .2941, r = - .5423
         c. MSR = 250, MSE = 25, $s_{Y|X}$ = 5.00

Difficulty level: Medium to Challenging

12.8.1: To check whether the data values used to estimate
the model parameters conform to the assumptions of
regression analysis, we use _____ and _____.

(a) scatter diagrams; stepwise regression
(b) scatter diagrams; residual analysis
(c) residual analysis; construct confidence intervals for
     the mean value of Y given X
(d) residual analysis; perform significance tests for β_1
(e) residual analysis; perform significance tests for ρ

Answer: (b)
Difficulty level: Medium

12.8.2: In a simple linear regression model, the assumption
of homoscedasticity is most closely associated with:

(a) residuals having a predictable pattern over time
(b) residuals being normally distributed
(c) residuals having a constant variance across values of X
(d) residuals that are not independently distributed
(e) none of these

Answer: (c)
Difficulty level: Medium

12.8.3: The violation of the assumption that the residuals
are independently and randomly distributed over time is
called:

(a) multicollinearity
(b) heteroscedasticity
(c) homoscedasticity
(d) autocorrelation
(e) nonlinearity

Answer: (d)
Difficulty level: Medium

Multiple Regression

13.1.1: In the multiple linear regression model, the error term ϵ is assumed to be a random variable with a mean or expected value of _____ and a variance _____.

(a) zero; that increases as the sample size increases
(b) zero; that increases as values of the independent variables increase
(c) zero; that is the same or constant for all values of the independent variables
(d) μ_i; that is the same or constant for all values of the independent variables
(e) μ; of σ^2

Answer: (c)
Difficulty level: Easy to Medium

13.1.2: In multiple regression analysis,

(a) the coefficient of determination must be larger than that for the simple linear regression model
(b) there must be only one independent variable
(c) there can be any number of dependent variables but only one independent variable
(d) it is more likely to obtain statistically significant variables than in simple linear regression
(e) none of these

Answer: (e)
Difficulty level: Easy to Medium

13.1.3: Multiple regression analysis is an estimation technique that is used to predict the effect on the:

(a) independent variable of a unit change in any dependent variable, holding the other dependent variables constant
(b) dependent variable of a unit change in any independent variable, holding the other independent variables constant
(c) dependent variable of a unit change in each of the independent variables
(d) independent variable of a unit change in each of the dependent variables
(e) none of these

Answer: (b)
Difficulty level: Easy to Medium

13.1.4: Multiple regression is the appropriate model to use:

(a)   when multiple forecasts are desired
(b)   when more than one dependent variable exists
(c)   when more than one simple linear model is desired
(d)   for predicting the value of an independent variable, whose value depends on several dependent variables
(e)   for predicting the value of a dependent variable, whose value depends on several independent variables

Answer: (e)
Difficulty level: Easy

13.1.5: Multiple regression analysis is widely used because:

(a)   it allows us to incorporate the separate effect of categorical or indicator variable on the value of a dependent variable
(b)   often we are interested in predicting the value of a dependent variable whose value depends on several variables
(c)   the assumptions upon which it is based are rarely violated
(d)   with it we can predict the separate effect on a dependent variable of unit changes in any independent variable, holding the other independent variables constant
(e)   more than one of these is correct

Answer: (e)
Difficulty level: Easy to Medium

13.1.6: Which of the following are assumptions of the multiple linear regression model?

I.   the X values for each independent variable are known, that is, are not random
II.   for each set of values for the independent variables, Y is normally and independently distributed
III.   for each set of values for the independent variables, the variance of Y given X_1, X_2, ... , X_k is the same.

(a)   I only
(b)   II only
(c)   III only
(d)   II and III only
(e)   all of these are assumptions of multiple linear regression analysis

Answer: (e)
Difficulty level: Easy to Medium

13.2.1: In a multiple regression setting with two independent variables, the method of least squares determines the two slope coefficients by finding the:

(a)   line which maximizes the coefficient of determination
(b)   line which minimizes the sum of squared deviations between the observed values of the dependent variable and the line
(c)   line which minimizes the sum of squared deviations between the observed values of the independent variables and the line
(d)   plane which minimizes the sum of squared deviations between the observed values of the dependent variable and the plane
(e)   plane which minimizes the sum of squared deviations between the observed values of the independent variables and the plane

Answer: (d)
Difficulty level: Easy to Medium

13.2.2: Given a sample multiple regression equation of the form Yhat = b_0 + b_1X_1 + b_2X_2, the term b_2 represents:

(a)   the percentage change in the predicted value of Y for a one percentage change in X_1, holding X_2 constant
(b)   the percentage change in the predicted value of Y for a one percentage change in X_2, holding X_1 constant
(c)   the change in the predicted value of Y for a unit change in X_1, holding X_2 constant
(d)   the change in the predicted value of Y for a unit change in X_2, holding X_1 constant
(e)   the proportion of total variability in Y that is accounted for or explained by the independent variables

Answer: (d)
Difficulty level: Easy to Medium

13.2.3: Given a sample multiple regression equation of the form Yhat = 5 + $3X_1$ - $2X_2$, the predicted value of Y, given that X_1 = 6 and X_2 = 7, is:

(a)   4
(b)   9
(c)   37
(d)   32
(e)   none of these

Answer: (b)
Difficulty level: Easy

13.3.1: In a multiple regression setting with *two* independent variables, which of the following equalities are true?

I.   SST = SSR + SSE
II. MSR = SSR/1
III. MSE = SSE/(n - 2)
IV. F = MSR/MSE

(a) I only
(b) I and IV only
(c) II and III only
(d) I, III, and IV only
(e) I, II, III, and IV

Answer: (b)
Difficulty level: Medium

13.3.2: A multiple regression analysis with sales (in dollars) of a product as the dependent variable Y, and advertising expenditure X_1 and market share X_2 as independent variables, was done. A random sample of 9 retail outlets yielded values for the regression and total sum of squares of SSR = 15 and SST = 24. The values for MSR and MSE are:

(a) 15 and 9
(b) 7.5 and 4.5
(c) 7.5 and 1.5
(d) 7.5 and 1.225
(e) cannot be determined with the information given

Answer: (c)
Difficulty level: Easy to Medium

13.3.3: A multiple regression analysis with sales (in dollars) of a product as the dependent variable Y, and advertising expenditure X_1 and market share X_2 as independent variables, was done. A random sample of 10 retail outlets yielded values for the regression and total sum of squares of SSR = 15 and SST = 24. The value of the standard error of the estimate is:

(a) 7.5
(b) 1.5
(c) 2.739
(d) 1.225
(e) cannot be determined with the information given

Answer: (d)
Difficulty level: Medium

13.4.1: As explanatory variables are added to a regression model, the coefficient of multiple determination:

(a)   must increase or stay the same
(b)   must decrease or stay the same
(c)   may increase, stay the same, or decrease
(d)   can increase to beyond one for significant variables
(e)   never changes

Answer: (a)
Difficulty level: Medium

13.4.2: As explanatory variables are added to a regression model, the *adjusted* coefficient of multiple determination:

(a)   must increase or stay the same
(b)   must decrease or stay the same
(c)   may increase, stay the same, or decrease
(d)   can increase to beyond one for significant variables
(e)   never changes

Answer: (c)
Difficulty level: Medium

13.4.3: The use of the adjusted R^2 value rather than the unadjusted R^2 value generally results in:

(a)   a smaller number
(b)   a larger number
(c)   the same number
(d)   a value of zero
(e)   a value near unity

Answer: (a)
Difficulty level: Easy

13.4.4: The term "adjusted" in the R^2 calculation, means:

(a)   Adjusted for sample size
(b)   Adjusted for degrees of freedom
(c)   Adjusted for the number of independent variables
(d)   More than one of these
(e)   None of these

Answer: (d)
Difficulty level: Medium

13.4.5: The coefficient of multiple determination:

(a)   indicates the proportion of total variability in the
      independent variables accounted for by the dependent
      variable
(b)   indicates the proportion of total variability in the
      dependent variable accounted for by any one of the
      independent variables
(c)   indicates the proportion of total variability in the
      independent variable accounted for by any one of the
      dependent variables
(d)   indicates the proportion of total variability in the
      dependent variable accounted for by all of the
      independent variables
(e)   tells us the effect on a dependent variable of a unit
      change in an independent variable, holding the other
      independent variables constant

Answer: (d)
Difficulty level: Easy to Medium

13.4.6: The *adjusted* coefficient of multiple determination
will be significantly higher than the coefficient of
multiple determination:

(a)   when the sample size is small
(b)   when the number of independent variables is large
(c)   when none of the independent variables are significant
(d)   when all of the independent variables are statistically
      significant
(e)   the adjusted coefficient of multiple determination is
      never larger than the coefficient of multiple
      determination

Answer: (e)
Difficulty level: Medium

13.4.7: In performing a multiple regression analysis, a
coefficient of multiple determination R^2 of .94 was found
for a multiple regression model. The adjusted R^2 for the
same model is:

(a)   greater than or equal to .94
(b)   less than or equal to .94
(c)   greater than one
(d)   one
(e)   unable to be determined with the information given

Answer: (b)
Difficulty level: Easy to Medium

13.4.8: The *adjusted* coefficient of multiple determination will be significantly lower than the coefficient of multiple determination:

I.   when the sample size is small
II.   when the number of independent variables is large
III. when all of the independent variables are statistically significant

(a)   I only
(b)   II only
(c)   III only
(d)   I and II only
(e)   I, II, and III

Answer: (d)
Difficulty level: Medium

13.4.9: A multiple regression analysis with annual profits of a pharmaceutical product as the dependent variable Y, and annual advertising expenditures X_1 and annual sales X_2 as the independent variables, was done. A random sample of 8 retail outlets yielded values for the coefficient of multiple determination and total sum of squares of .80 and 600. From this information, the standard error of the estimate is:

(a)   480
(b)   120
(c)   4.899
(d)   4.472
(e)   unable to be determined with the information given

Answer: (c)
Difficulty level: Challenging

13.4.10: A multiple regression analysis with annual profits of a pharmaceutical product as the dependent variable Y, and annual advertising expenditures X_1 and annual sales X_2 as the independent variables, was done. A random sample of size 10 yielded values for the regression and total sum of squares of 400 and 500. The standard error of the estimate is:

(a)   100
(b)   14.286
(c)   3.780
(d)   3.536
(e)   unable to be determined with the information given

Answer: (c)
Difficulty level: Challenging

13.5.1: A financial analyst wants to develop a multiple regression model to predict the average annual rate of return on stocks (Y, expressed as a percent) from their price-earnings ratio (X_1) and a measure of risk known as the stock's beta (X_2). A sample of 10 stocks was taken, yielding the following regression equation:

$$\hat{Y} = -16.9631 + .4907X_1 + 23.0361X_2,$$
$$\qquad\qquad\quad (.3281) \quad\; (5.4613)$$

with standard errors reported in parentheses. Additional regression information is: $S_{Y|X} = 1.619$, $R^2 = .723$, SST = 66.18, MSR = 23.910, and MSE = 2.622. The predicted annual rate of return for a stock with a price-earnings ratio of 12 and a beta value of 1.2 is:

(a)   16.57 percent
(b)   33.521 percent
(c)   -16.963 percent
(d)   -11.075 percent
(e)   none of these

Answer: (a)
Difficulty level: Easy to Medium

13.5.2: A financial analyst wants to develop a multiple regression model to predict the average annual rate of return on stocks (Y, expressed as a percent) from their price-earnings ratio (X_1) and a measure of risk known as the stock's beta (X_2). A sample of 10 stocks was taken, yielding the following regression equation:

$$\hat{Y} = -16.9631 + .4907X_1 + 23.0361X_2,$$
$$\qquad\qquad\quad (.3281) \quad\; (5.4613)$$

with standard errors reported in parentheses. Additional regression information is: $S_{Y|X} = 1.619$, $R^2 = .723$, SST = 66.18, MSR = 23.910, and MSE = 2.622. At $\alpha = .01$, we should:

(a)   conclude that both variables are significant
(b)   conclude that both variables are insignificant
(c)   conclude that X_1 is significant and X_2 is insignificant
(d)   conclude that X_1 is insignificant and X_2 is significant
(e)   conclude that the regression explains less than 50% of the total variability in stock annual rates of returns

Answer: (d)
Difficulty level: Medium

13.5.3: A prospective home buyer hired a consultant to develop a multiple regression model to predict mortgage interest rates (Y, expressed as a percent) by using the annual yield on long-term Treasury bonds (X_1, expressed as a percent) and the type of mortgage (X_2, variable rate = 1, fixed rate = 0). A sample of 20 mortgages was taken, yielding the regression equation:

$$\hat{Y} = 5.9577 + .8037X_1 - 1.9386X_2,$$
$$\phantom{\hat{Y} = 5.9577 +}(.1570) \quad (.5715)$$

with standard errors reported in parentheses. Additional regression information is: $S_{Y|X} = 1.207$, $R^2 = .643$, SST = 69.35, MSR = 22.30, and MSE = 1.46. The predicted mortgage interest rate on a variable-rate mortgage, given an annual long-term Treasury bond yield of 10%, is:

(a)   18.01 percent
(b)   12.06 percent
(c)   5.9577 percent
(d)   13.995 percent
(e)   none of these

Answer: (b)
Difficulty level: Easy to Medium

13.5.4: A prospective home buyer hired a consultant to develop a multiple regression model to predict mortgage interest rates (Y, expressed as a percent) by using the annual yield on long-term Treasury bonds (X_1, expressed as a percent) and the type of mortgage (X_2, variable rate = 1, fixed rate = 0). A sample of 20 mortgages was taken, yielding the regression equation:

$$\hat{Y} = 5.9577 + .8037X_1 - 1.9386X_2,$$
$$\phantom{\hat{Y} = 5.9577 +}(.1570) \quad (.5715)$$

with standard errors reported in parentheses. Additional regression information is: $S_{Y|X} = 1.207$, $R^2 = .643$, SST = 69.35, MSR = 22.30, and MSE = 1.46. At $\alpha = .01$, we should:

(a)   conclude that both variables are significant
(b)   conclude that both variables are insignificant
(c)   conclude that X_1 is significant and X_2 is insignificant
(d)   conclude that X_1 is insignificant and X_2 is significant
(e)   conclude that the regression explains less than 50% of the total variability in mortgage interest rates

Answer: (a)
Difficulty level: Medium

13.5.5: The number of annual employee absences (Y, measured by an index) is in part determined by the employee's job tenure (X_1, measured in number of months with the company) and the employee's job satisfaction (X_2, measured by an index). For a sample of 10 workers, the estimated regression equation is:

$$\hat{Y} = 51.5793 - .4008X_1 - .1947X_2,$$
$$\quad\quad\quad (.1175) \quad (.0816)$$

with standard errors reported in parentheses. Additional regression information is: $S_{Y|X} = 2.465$, $R^2 = .861$, SST = 305.4, MSR = 131.42, and MSE = 6.077. The predicted value of Y given $X_1 = 40$ and $X_2 = 50$, is:

(a)   51.5793
(b)   -25.765
(c)   25.812
(d)   77.346
(e)   none of these

Answer: (c)
Difficulty level: Easy to Medium

13.5.6: The number of annual employee absences (Y, measured by an index) is in part determined by the employee's job tenure (X_1, measured in number of months with the company) and the employee's job satisfaction (X_2, measured by an index). For a sample of 10 workers, the estimated regression equation is:

$$\hat{Y} = 51.5793 - .4008X_1 - .1947X_2,$$
$$\quad\quad\quad (.1175) \quad (.0816)$$

with standard errors reported in parentheses. Additional regression information is: $S_{Y|X} = 2.465$, $R^2 = .861$, SST = 305.4, MSR = 131.42, and MSE = 6.077. At $\alpha = .01$, we should:

(a)   conclude that both variables are significant
(b)   conclude that both variables are insignificant
(c)   conclude that X_1 is significant and X_2 is insignificant
(d)   conclude that X_1 is insignificant and X_2 is significant
(e)   conclude that the regression explains less than 50% of
       the total variability in annual employee absences

Answer: (b)
Difficulty level: Easy

13.5.7: An individual's annual wages (Y, measured in thousands of dollars) is in part determined by their experience (X_1, measured in number of months with the company) and education (X_2, College degree = 1, No College Degree = 0). For a sample of 20 workers, the estimated regression equation is:

$$\hat{Y} = .0075 + .4852X_1 + 5.532X_2,$$
$$\phantom{\hat{Y} = .0075 +} (.0293) \quad (0.617)$$

with standard errors reported in parentheses. Additional regression information is: $S_{Y|X} = 1.374$, $R^2 = .951$, SST = 655.5, MSR = 311.71, and MSE = 1.888. The predicted annual wages for an individual with a college degree and 24 months of job experience (in dollars) is:

(a) $7.50
(b)   approximately $17.18
(c)   approximately $17,184.0
(d)   approximately $17,176.8
(e)   unable to be determined with the information given

Answer: (c)
Difficulty level: Medium

13.5.8: An individual's annual wages (Y, measured in thousands of dollars) is in part determined by their experience (X_1, measured in number of months with the company) and education (X_2, College degree = 1, No College Degree = 0). For a sample of 20 workers, the estimated regression equation is:

$$\hat{Y} = .0075 + .4852X_1 + 5.532X_2,$$
$$\phantom{\hat{Y} = .0075 +} (.0293) \quad (0.617)$$

with standard errors reported in parentheses. Additional regression information is: $S_{Y|X} = 1.374$, $R^2 = .951$, SST = 655.5, MSR = 311.71, and MSE = 1.888. At $\alpha = .05$, we should:

(a)   conclude that both variables are significant
(b)   conclude that both variables are insignificant
(c)   conclude that X_1 is significant and X_2 is insignificant
(d)   conclude that X_1 is insignificant and X_2 is significant
(e)   conclude that the regression explains less than 50% of the total variability in annual wages for individuals

Answer: (a)
Difficulty level: Medium

13.5.9: A student's score on the CPA exam (Y, measured by the percentage of correct responses) is in part determined by the student's cumulative GPA (X_1) and the student's experience in accounting (X_2, measured in number of months working in an accounting-related job). For a sample of 10 students, the estimated regression equation is:

$$\hat{Y} = -21.5907 + 21.6822X_1 + 2.9818X_2$$
$$(5.5363) \quad (.7098)$$

with standard errors reported in parentheses. Additional regression information is: $S_{Y|X} = 6.135$, $R^2 = .8295$, SST = 1544.93, MSR = 640.75, and MSE = 37.633. The predicted score on the CPA exam for a student with a GPA of 3.0 and three months of job experience (in percent) is:

(a)   -21.591
(b)   20.710
(c)   91.880
(d)   70.290
(e)   52.401

Answer: (e)
Difficulty level: Easy to Medium

13.5.10: A student's score on the CPA exam (Y, measured by the percentage of correct responses) is in part determined by the student's cumulative GPA (X_1) and the student's experience in accounting (X_2, measured in number of months working in an accounting-related job). For a sample of 10 students, the estimated regression equation is:

$$\hat{Y} = -21.5907 + 21.6822X_1 + 2.9818X_2$$
$$(5.5363) \quad (.7098)$$

with standard errors reported in parentheses. Additional regression information is: $S_{Y|X} = 6.135$, $R^2 = .8295$, SST = 1544.93, MSR = 640.75, and MSE = 37.633. At $\alpha = .01$, we should conclude that:

(a)   both variables are significant
(b)   both variables are insignificant
(c) X_1 is significant and X_2 is insignificant
(d) X_1 is insignificant and X_2 is significant
(e)   the regression explains less than 50% of the total variability in student scores on the CPA exam

Answer: (a)
Difficulty level: Medium

13.5.11: A marketing manager with a large corporation hires a consultant to develop a multiple regression model to predict sales (Y, measured in thousands of dollars) from advertising expenditures (X_1, measured in thousands of dollars) and market share (X_2, given as a percentage). A sample of 10 periods was taken, yielding the sample regression equation:

$$\hat{Y} = 1.7071 + .9125X_1 + .2213X_2 ,$$
$$\qquad\qquad (.4612) \quad (.0682)$$

with standard errors reported in parentheses. Additional regression information is: $S_{Y|X} = 1.152$, $R^2 = .615$, SST = 24.14, MSR = 7.429, and MSE = 1.327. The predicted sales for a given level of $5000 in advertising expenses and a market share of 25% (in dollars) is:

(a) $11.80
(b) $1,707.10
(c) $10,095.50
(d) $11,802.10
(e) $11,802,100

Answer: (d)
Difficulty level: Easy to Medium

13.5.12: A marketing manager with a large corporation hires a consultant to develop a multiple regression model to predict sales (Y, measured in thousands of dollars) from advertising expenditures (X_1, measured in thousands of dollars) and market share (X_2, given as a percentage). A sample of 10 periods was taken, yielding the sample regression equation:

$$\hat{Y} = 1.7071 + .9125X_1 + .2213X_2 ,$$
$$\qquad\qquad (.4612) \quad (.0682)$$

with standard errors reported in parentheses. Additional regression information is: $S_{Y|X} = 1.152$, $R^2 = .615$, SST = 24.14, MSR = 7.429, and MSE = 1.327. At $\alpha = .05$, we should:

(a)   conclude that both variables are significant
(b)   conclude that both variables are insignificant
(c)   conclude that X_1 is significant and X_2 is insignificant
(d)   conclude that X_1 is insignificant and X_2 is significant
(e)   conclude that the regression explains less than 50% of the total variability in the dependent variable sales

Answer: (d)
Difficulty level: Medium

13.6.1: In explaining annual incomes of United States citizens, which of the following explanatory variables is best captured with an indicator or dummy variable?

(a)   age
(b)   education level (number of years completed)
(c)   gender
(d)   occupation
(e)   none of these

Answer: (c)
Difficulty level: Easy to Medium

13.6.2: In explaining average starting salaries for law school graduates, which of the following explanatory variables is best captured with an indicator or dummy variable?

(a)   student's age upon graduation from law school
(b)   whether student passed their state bar exam
(c)   student grade point average in law school
(d)   name of law school student graduated from
(e)   all of these

Answer: (b)
Difficulty level: Easy to Medium

13.6.3: In explaining annual profits for pharmaceutical companies, which of the following explanatory variables is best captured with an indicator or dummy variable?

(a)   annual sales of the pharmaceutical company
(b)   annual R&D expenditures of the pharmaceutical company
(c)   number of years the company has been in existence
(d)   market share of the pharmaceutical company
(e)   none of these

Answer: (e)
Difficulty level: Easy to Medium

13.6.4: In explaining home mortgage rates, which of the following explanatory variables is best captured with an indicator or dummy variable?

(a)   Federal funds interest rates
(b)   Treasury-Bill bond annual yields
(c)   whether the mortgage is fixed-rate or variable-rate
(d)   unemployment rate among construction workers
(e)   none of these

Answer: (c)
Difficulty level: Easy to Medium

13.7.1: The standard error of the estimate will fail to be a consistent estimator of the population standard deviation of the dependent variable for given values of the independent variables in the presence of:

(a)   linearity
(b)   nonlinearity
(c)   heteroscedasticity
(d)   autocorrelation
(e)   multicollinearity

Answer: (c)
Difficulty level: Medium

13.7.2: If the null hypothesis that all of the β_i's equal zero is false, then which of the following statements are true?

I.   the numerator of the F statistic provides an unbiased estimate of the common population variance σ^2
II.   the denominator of the F statistic provides an unbiased estimate of the common population variance σ^2
III.   we can be confident that all of the independent variables are statistically significant

(a)   Only statement I is true
(b)   Only statement II is true
(c)   Only statement III is true
(d)   Only statements II and III are true
(e)   Statements I, II, and III are all true

Answer: (b)
Difficulty level: Medium

13.8.1: Heteroscedasticity means that:

(a)   the independent variables are highly correlated with one another
(b)   the distribution of the dependent variable has a variance that is constant for all combinations of the independent variables
(c)   the distribution of the dependent variable has a variance that is not constant for all combinations of the independent variables
(d)   the error terms are correlated with one another
(e)   there must be a curvilinear relationship between the dependent and independent variables

Answer: (c)
Difficulty level: Easy to Medium

13.8.2: Multicollinearity means that:

(a)   the independent variables are highly correlated with
      one another
(b)   the distribution of the dependent variable has a
      variance that is constant for all combinations of the
      independent variables
(c)   the distribution of the dependent variable has a
      variance that is not constant for all combinations of
      the independent variables
(d)   the error terms are correlated with one another
(e)   there must be a curvilinear relationship between the
      dependent and independent variables

Answer: (a)
Difficulty level: Easy to Medium

13.8.3: Overspecification or "overfitting" in a regression
problem occurs due to:

(a)   too many significant independent variables being
      included in the final model
(b)   too many independent variables, not significantly
      different from zero, being included in the model
(c)   using too large a sample
(d)   using a quadratic equation when a linear equation
      should be used
(e)   more than one of these is correct

Answer: (b)
Difficulty level: Easy to Medium

13.8.4: Which of the following is an advantage for looking
at the correlation matrix?

(a)   May show potential multicollinearity
(b)   May show potentially insignificant variables
(c)   Shows the correlation coefficients between all possible
      pairs of variables
(d)   Shows potential multicollinearity, potential
      insignificant variables, and the correlation
      coefficients between all possible pairs of variables
(e)   None of these

Answer: (d)
Difficulty level: Easy to Medium

13.8.5: One of the best ways to determine whether some of the basic assumptions of the regression procedure have been violated is to look at:

(a)   the value of R^2
(b)   the size of SSE
(c)   a graph of the residuals plotted against each of the independent variables
(d)   a random numbers table
(e)   the number of independent variables in the model

Answer: (c)
Difficulty level: Easy to Medium

13.8.6: In a correctly specified linear regression model, residual plots of the independent variables should have the points:
(a)   in a straight line
(b)   randomly distributed above and below zero
(c)   all below zero
(d)   all above zero
(e)   in a straight line and all below zero

Answer: (b)
Difficulty level: Medium

13.8.7: The specific problem introduced by the presence of multicollinearity in the data set manifests itself through:

(a)   inflated s_{b_j} values
(b)   deflated t-values for some of the independent variables
(c)   failure to detect significant relationships between Y and some of the X's
(d)   large differences between R^2 and adjusted R^2
(e)   more than one of these is correct

Answer: (e)
Difficulty level: Medium

13.8.8: The appropriate test (or statistical technique) for detecting autocorrelation is:

(a)   an F test
(b)   a χ^2 test
(c)   the Durbin-Watson test
(d)   stepwise regression
(e)   none of these

Answer: (c)
Difficulty level: Easy to Medium

13.9.1: In stepwise multiple regression, the regression equation is constructed by:

(a)   entering all dependent variables at once.
(b)   entering all independent variables at once.
(c)   entering the independent variables one at a time.
(d)   more than one of these
(e)   none of these

Answer: (d)
Difficulty level: Easy to Medium

13.9.2: The X variable that is entered on each step of the stepwise procedure is:

(a)   the one having the largest values.
(b)   the one that does the best job of explaining the remaining variation in Y.
(c)   the one that does the best job in eliminating the prediction errors of the previous equation.
(d)   the one that most correctly specifies the model.
(e)   more than one of these is correct.

Answer: (e)
Difficulty level: Easy to Medium

13.9.3: In stepwise multiple regression, we would generally continue to bring X's into the equation until:

(a)   the value for adjusted R^2 is sufficiently high
(b)   the value for adjusted R^2 is sufficiently low
(c)   the value for adjusted R^2 is positive
(d)   the model is correctly specified
(e)   all of the originally defined set of X's are in the model

Answer: (a)
Difficulty level: Medium

13.11.1: For the regression equation $Y = 20 + 15X_1 + .001X_2$, hypothesis tests $H_0: B_1 = 0$ and $H_0: B_2 = 0$ were performed. It was concluded that $B_1 = 0$ but that $B_2 \neq 0$. Can you justify these results?

Answer: Yes. The ratios of an estimated coefficient to the estimated standard error of the respective slopes are what determine the t-values. The standard error of B_1 may have been very large relative to the estimated coefficient while for B_2 it may have been very small.
Difficulty level: Medium

Time Series Analysis

14.1.1: Much of the historical data available to economists and people in business has been recorded at regular time intervals to form a:

(a) business forecasting model
(b) cross section
(c) time series
(d) long-term trend
(e) short-term trend

Answer: (c)
Difficulty level: Easy

14.1.2: Monthly and quarterly figures for gross national product, inflation, and unemployment are all examples of:

(a) business forecasting models
(b) cyclical variation
(c) seasonal variation
(d) time series data
(e) long-term trends

Answer: (d)
Difficulty level: Easy

14.1.3: Forecasting future level of company employment might be useful for determining:

(a) how many company uniforms to order
(b) whether to plan an expansion of the company cafeteria
(c) company costs of worker retirement and pension plans
(d) budgeting insurance premium costs
(e) all of these

Answer: (e)
Difficulty level: Easy

14.1.4: Historical data:

(a) are used to estimate probabilities
(b) are used in regression and correlation problems
(c) have been recorded at regular time intervals
(d) are used to forecast values of desired variables
(e) all of these

Answer: (e)
Difficulty level: Easy

14.1.5: Time series data are most often graphed using:

(a)   a histogram
(b)   a frequency distribution
(c)   a time series plot or line chart
(d)   a bar chart
(e)   none of these

Answer: (b)
Difficulty level: Easy

14.1.6: Time series analysis consists of breaking down a time series into its components and analyzing these components so that:

(a)   we can construct a trend
(b)   we can graph the pattern
(c)   concrete, quantitative statements about the patterns the data reveal can be made
(d)   concrete, qualitative statements about the patterns which show in the graph can be made
(e)   none of these

Answer: (c)
Difficulty level: Easy

14.2.1: The components of time series do not include:

(a)   short-term trend
(b)   long-term trend
(c)   seasonal variations
(d)   cyclical variations
(e)   irregular variations

Answer: (a)
Difficulty level: Easy

14.2.2: In time series, the long-term movements of the series that can be characterized by steady or only slightly variable rates of change are:

(a) rare
(b) trends
(c) referred to as seasonal variations
(d) referred to as cyclical variations
(e) referred to as irregular variations

Answer: (b)
Difficulty level: Easy

14.2.3: In time series, those variations that occur rather predictably at a particular time each year are:

(a) rare
(b) trends
(c) referred to as seasonal variations
(d) referred to as cyclical variations
(e) referred to as irregular variations

Answer: (c)
Difficulty level: Easy

14.2.4: Movements in a time series that are recurrent but that occur in patterns of longer than one year are:

(a) rare
(b) trends
(c) referred to as seasonal variations
(d) referred to as cyclical variations
(e) referred to as irregular variations

Answer: (d)
Difficulty level: Easy

14.2.5: Variations that cannot be predicted using historical data and which are not periodic in nature because they are caused by weather, wars, strikes, and elections are:

(a) rare
(b) trends
(c) referred to as seasonal variations
(d) referred to as cyclical variations
(e) referred to as irregular variations

Answer: (e)
Difficulty level: Easy

14.2.6: Irregular variations:

(a)   Are movements left in a time series when trends, seasonal, and cyclical variations have been identified
(b)   Cannot be predicted by using historical data
(c)   Are not periodic in nature
(d)   Are caused by such factors as weather, wars, strikes, and government legislation
(e)   All of these

Answer: (e)
Difficulty level: Easy to Medium

14.2.7: Which of the following statements is true?

(a) Trends in time series are the long term movements of
     the series that can be characterized by steady rates of
     change.
(b) Seasonal variations in a time series are those
     variations that occur rather predictably over a period
     of years.
(c) Cyclical variations in a time series are those
     variations that occur rather predictably at a
     particular time each year.
(d) Irregular variations is a mix of trend, seasonal, and
     cyclical variations.
(e) All of these.

Answer: (a)
Difficulty level: Easy to Medium

14.2.8: Seasonal variations can:

(a) only be found in quarterly data
(b) only be found in monthly data
(c) cannot be found in weekly data
(d) cannot be found in daily data
(e) can be found in quarterly, monthly, weekly, or daily
     data

Answer: (e)
Difficulty level: Medium

14.2.9: Which of the following statements is true?

(a) The multiplicative time series model assumes that the
     seasonal variations produce the same amount of change.
(b) The additive time series model assumes that seasonal
     variations produce the same percentage change each time
     a season arrives.
(c) Seasonal variations seem to be constant percentages
     through the years rather than constant amounts.
(d) All of these statements are true
(e) Irregular variations are movements in a time series
     that occur in cycles of longer than a year and are not
     affected by events like war, strikes, or the weather

Answer: (c)
Difficulty level: Easy to Medium

14.2.10: The table below indicates the number of mergers that took place in an industry over a 19-year period. What type of trend might best be fit to this time series?

Year	Mergers	Year	Mergers	Year	Mergers
1	23	8	64	14	150
2	23	9	47	15	165
3	31	10	96	16	192
4	23	11	125	17	210
5	32	12	140	18	250
6	32	13	160	19	300
7	42				

(a)   a linear short-term trend
(b)   a linear long-term trend
(c)   a curved short-term trend
(d)   a curved long-term trend
(e)   any of these would provide an adequate fit to the data

Answer: (d)
Difficulty level: Easy to Medium

14.3.1: Smoothing techniques include:

(a)   moving average only
(b)   autoregression only
(c)   exponential smoothing only
(d)   moving average and exponential smoothing
(e)   moving average and autoregression

Answer: (d)
Difficulty level: Easy

14.3.2: When two or more consecutive values in the series are averaged and the computed value replaces one of the values averaged, it is done by:

(a)   moving average
(b)   average smoothing
(c)   mean smoothing
(d)   trend average
(e)   exponential smoothing

Answer: (a)
Difficulty level: Easy to Medium

14.3.3: A problem with using a moving average is that:

(a)   the smoothed values contain more variation that
      original values
(b)   the smoothed values contain less variation that
      original values
(c)   the moving average cannot be calculated for all of the
      periods in the time series
(d)   it requires the use of discrete data
(e)   it requires the use of categorical data

Answer: (c)
Difficulty level: Easy to Medium

14.3.4: If in the exponential smoothing model the smoothing
value constant α is large, say, $\alpha = .95$, then S_t:

(a)   will equal 1
(b)   will equal 0
(c)   will be composed primarily of the historical value of
      the time series
(d)   will be composed primarily of the current value of the
      time series
(e)   will be composed primarily of the future value of the
      time series

Answer: (d)
Difficulty level: Easy to Medium

14.3.5: Which of the following is true about exponential
smoothing?

(a)   The exponentially smoothed value for period t is a
      weighted sum of all previous values in the time series
      back to time t = 1.
(b)   If the smoothing constant is large, then the
      exponentially smoothed value for period t will be
      dependent on the historical values of the series.
(c)   If the analyst wants the smoothed series to follow the
      actual series quite closely, he/she should use a small
      smoothing constant.
(d)   Forecasters commonly use a smoothing constant between
      0.3 and 0.7.
(e)   None of these.

Answer: (a)
Difficulty level: Easy to Medium

14.3.5: The forecasting method that is appropriate when the time series has no significant trend, cyclical, or seasonal effect is:

(a) exponential smoothing
(b) moving averages
(c) using regression analysis to fit a trend line
(d) qualitative forecasting methods
(e) none of these

Answer: (b)
Difficulty level: Medium

14.3.6: Once the original data of a time series have been smoothed, the results indicate that the series might contain elements of:

I.   trend
II.   seasonal variations
III. cyclical variations
IV.   irregular fluctuations

(a) I only
(b) IV only
(c) II and III only
(d) I and IV only
(e) I, II, and III only

Answer: (e)
Difficulty level: Easy to Medium

14.4.1: Describe in words a procedure for forecasting the future period's value of a time series.

Answer: If we wish to forecast a future period's value of
            the time series, we can find the forecast using
            the following four step procedure.
            STEP 1:   Project a trend value (T)
            STEP 2:   Multiply the trend value (T) by a
                      seasonal index (S) that measures the
                      typical deviation of the time series
                      from the trend for that season
            STEP 3:   Multiply the value obtained in step 2 by
                      an index that measures the cyclical
                      variation expected for that period (C)
            STEP 4:   Obtain the forecasted value by computing
                      the value of T x S x C
Difficulty level: Medium

14.4.2: When the original data in a time series have been smoothed, the results often uncover evidence that the time series contains elements of:

(a)   trends only
(b)   seasonal variations only
(c)   cyclical variations only
(d)   seasonal and cyclical variations only
(e)   trends, seasonal, and cyclical variations

Answer: (e)
Difficulty level: Easy to Medium

14.4.3: The most common method for finding a trend line is:

(a)   to use a moving average
(b)   to use exponential smoothing
(c)   to fit a least squares (regression) line to the data
(d)   to fit a polynomial trend line to the data
(e)   none of these

Answer: (c)
Difficulty level: Easy

14.5.1: The least-squares line is the most common method used to:

(a)   analyze seasonal variations
(b)   smooth cyclical variations
(c)   correlate current periods with historical data
(d)   find a straight line to represent seasonal variations
(e)   find a trend line

Answer: (e)
Difficulty level: Easy to Medium

14.5.2: The least-squares method of fitting a line to a set of points:

(a)   is most often used for cyclical variation analysis
(b)   is most often used for seasonal variation analysis
(c)   gives the same weight to all fluctuations
(d)   gives heavy weight to the smaller fluctuations
(e)   gives heavy weight to the larger fluctuations

Answer: (e)
Difficulty level: Easy to Medium

14.6.1: The *ratio-to-moving-average* method involves:

(a)   breaking down a time series into its components
(b)   averaging two or more consecutive values of a time
      series
(c)   comparing each season's actual value with a yearly
      moving average to obtain an index
(d)   using the formula: $S_t = \alpha Y_t + (1 - \alpha)Y_{t-1}$ to obtain an
      exponentially smoothed value for period t
(e)   using the formula: forecasted value = T x S x C to
      obtain forecasted values of the time series for time
      period t

Answer: (c)
Difficulty level: Easy to Medium

14.6.2: The values in the accompanying table show the
percentage of students dropping a particular class over the
past three years at a school that is on a three-quarter
academic year. Use the *ratio-to-moving-average* method to
compute the following.

YEAR	Fall	QUARTER Winter	Spring
Two Years Ago	5	4	8
Last Year	6	4	10
This Year	9	8	12

a.   What is the index of drops for the fall quarter?
b.   What is the index of drops for the winter quarter?
c.   What is the index of drops for the spring quarter?

Answer: a.   99.0
           b.   70.4
           c.   130.6
Difficulty level: Medium

14.6.3: The means of four quarters are 100, 125, 125, and
130, respectively. The seasonal index for the first quarter
is:

(a)   80.00
(b)   83.33
(c)   104.17
(d)   108.33
(e)   unable to be determined with the information given

Answer: (b)
Difficulty level: Easy

14.6.4: The means of four quarters are 100, 125, 125, and 130, respectively. The seasonal index for the fourth quarter is:

(a)   80.00
(b)   83.33
(c)   104.17
(d)   108.33
(e)   unable to be determined with the information given

Answer: (d)
Difficulty level: Easy

14.6.5: Yo Yo Yogurt, Inc., has the following quarterly sales figures for their frozen yogurt for the years 1983 - 1987. These figures are presented below:

| | **QUARTERLY SALES (in thousands)** | | | |
YEAR	**I**	**II**	**III**	**IV**
1983	10	18	53	9
1984	12	23	70	10
1985	16	25	67	15
1986	20	50	100	25
1987	30	76	110	28

Construct the quarterly seasonal index for Yo Yo Yogurt. Does a seasonal pattern exist? Explain.

Answer: The seasonal indices for the four quarters are:
    Quarter I   45.9
    Quarter II   100.1
    Quarter III   208.6
    Quarter IV   45.4

    Yes, there is a seasonal pattern. Quarters I & IV are below normal, Quarter II is normal, while Quarter III is very much above normal relative to average yogurt sales. Frozen yogurt is much more popular during the summer months of the year.
Difficulty level: Medium

14.6.6: The term "seasonally adjusted" data means:

(a)   The seasonal effect had been added to the data.
(b)   The seasonal effect had not been measured in that data.
(c)   The irregular variations are eliminated.
(d)   The seasonal effect had been removed from the data.
(e)   None of these.

Answer: (d)
Difficulty level: Easy to Medium

14.7.1: Long-Haul Trucking, Inc., hired a consultant to determine whether the fuel consumption at the company contained seasonal components. Based on a sample of 20 quarters (up to the previous year), data analysis systems were used to obtain the following estimated regression equation:

$$\hat{Y} = 61.5 + 4.7t + 36.3S_2 + 74.6S_3 + 26.8S_4 ,$$

where the subscripts '2', '3', and '4' indicate the spring, summer, and fall seasons, respectively. The trend coefficient of 4.7 suggests that, other things constant:

(a) the fuel consumption has been falling at an annual rate of 4700 gallons
(b) the fuel consumption has been rising at an annual rate of 4700 gallons
(c) the fuel consumption has been rising at a quarterly rate of 4700 gallons
(d) the fuel consumption has been rising at a monthly rate of 4700 gallons
(e) none of these

Answer: (c)
Difficulty level: Easy to Medium

14.7.2: Long-Haul Trucking, Inc., hired a consultant to determine whether the fuel consumption at the company contained seasonal components. Based on a sample of 20 quarters (up to the previous year), data analysis systems were used to obtain the following estimated regression equation:

$$\hat{Y} = 61.5 + 4.7t + 36.3S_2 + 74.6S_3 + 26.8S_4 ,$$

where the subscripts '2', '3', and '4' indicate the spring, summer, and fall seasons, respectively. The forecasted value for the fall of this year (using t = 24) is:

(a) 61,500 gallons
(b) 174,300 gallons
(c) 26,800 gallons
(d) 201,100 gallons
(e) 248,900 gallons

Answer: (d)
Difficulty level: Easy to Medium

14.8.1: Cyclical variations are identified through:

(a)   indexes
(b)   least squares
(c)   moving averages
(d)   isolating the other components
(e)   all of these

Answer: (d)
Difficulty level: Easy to Medium

14.8.2: In the multiplicative model where Y = T x S x C x I, one may isolate cyclical fluctuations by:

(a)   dividing through by C
(b)   multiplying by C
(c)   dividing through by T x S and assuming I averages out
(d)   dividing through by T x S x C
(e)   dividing through by Y

Answer: (c)
Difficulty level: Easy to Medium

14.9.1: The January through December seasonal indices for Nena Apparel are: 80, 82, 92, 95, 100, 102, 104, 108, 110, 120, 107, 100. If the annual forecast for the coming year for Nena is $600,000, the forecasted value for January is:

(a)   about $100
(b)   about $40,000
(c)   about $48,723
(d)   about $50,000
(e)   about $60,000

Answer: (b)
Difficulty level: Easy to Medium

14.9.2: The January through December seasonal indices for Nena Apparel are: 80, 82, 92, 95, 100, 102, 104, 108, 110, 120, 107, 100. If the annual forecast for the coming year for Nena is $500,000, the forecasted value for March is:

(a)   about $100
(b)   about $38,330
(c)   about $41,667
(d)   about $50,000
(e)   about $55,830

Answer: (b)
Difficulty level: Easy to Medium

14.9.3: A financial manager wanted to compute a forecast of revenues for his firm. He has the data of previous revenues for the ten years prior to 1976 and calculates a trend equation Yhat = 10 + .8t (in millions of dollars). The quarterly revenue indices were: 78, 93, 106, 123, respectively. The projection is to be based only on the trend and seasonal data. The forecasted value of firm revenues for the first quarter in 1976 (t = 41) is:

(a) $32.80
(b) $42.80
(c) $42,800
(d) $42,800,000
(e) unable to be determined with the information given

Answer: (a)
Difficulty level: Easy to Medium

14.9.4: The Doggie Groomer serviced 350 dogs during the past month and wanted to forecast the number of groomings it could expect for this month. The seasonal index for the past month was 105 and for the current month it is 102. How many dogs could they expect to service this month?

(a) 318
(b) 321
(c) 325
(d) 340
(e) 360

Answer: (d)
Difficulty level: Easy to Medium

14.9.5: Annual billings for a large investment consulting firm were $79,852 in March, and the March seasonal index for this firm's billings is 107.

a. What is the seasonally adjusted March billing figure?
b. What would be the expected annual billings based on the March figure?

Answer: a. $74,628
           b. $895,536
Difficulty level: Easy to Medium

14.10.1: Construction employment levels are known to vary during the year, with the level of employment being highest during the summer. A model designed to forecast construction employment should therefore use:

(a)   a moving average process
(b)   exponential smoothing
(c)   an autoregressive model
(d)   seasonal dummy variables
(e)   a time trend

Answer: (d)
Difficulty level: Medium

14.11.1: When the assumption of independence between observations is violated in least squares regression, the effects of autocorrelation are:

(a)   the least squares estimators of the regression
      coefficients are biased
(b)   the least squares estimators of the regression
      coefficients are biased and inefficient
(c)   the least squares estimators of the regression
      coefficients are inefficient
(d)   the least squares estimators of the regression
      coefficient are biased and inconsistent
(e)   the least squares estimators of the regression
      coefficient are biased, inefficient, and inconsistent

Answer: (c)
Difficulty level: Medium

14.11.2: In typical business applications, the null and alternative hypotheses for the Durbin-Watson test are:

(a) H_0: $\rho = 0$; H_1: $\rho \neq 0$
(b) H_0: $\rho \geq 0$; H_1: $\rho < 0$
(c) H_0: $\rho \leq 0$; H_1: $\rho > 0$
(d) H_0: $\rho \neq 0$; H_1: $\rho = 0$
(e)   none of these

Answer: (c)
Difficulty level: Easy to Medium

14.11.3: The test for the significance of first-order autocorrelation in a regression model, when least squares estimation is used with time series data, is called:

(a)   the Sign test
(b)   the Wilcoxon signed rank test
(c)   the Mann-Whitney test
(d)   the Durbin-Watson test
(e)   an F test

Answer: (d)
Difficulty level: Easy

14.11.4: The range of possible values for the Durbin-Watson test statistic is:

(a)   0 to 2
(b)   2 to 4
(c)   0 to 4
(d)   -4 to 0
(e)   -4 to 4

Answer: (c)
Difficulty level: Easy to Medium

14.11.5: In the Durbin-Watson test the effect of positive autocorrelation is to _____ the _____.

(a)   reduce, numerator
(b)   reduce, denominator
(c)   increase, numerator
(d)   increase, denominator
(e)   increase, numerator and denominator

Answer: (a)
Difficulty level: Easy to Medium

14.11.6: Given n = 30, two independent variables, a null hypothesis of no positive autocorrelation, and α = .05, the null hypothesis will be rejected with a Durbin-Watson test statistic of:

(a) 1.01
(b) 1.36
(c) 1.68
(d) 2.50
(e) 2.75

Answer: (e)
Difficulty level: Easy to Medium

14.11.7: Given n = 30, two independent variables, and α = .05, the null hypothesis of no positive autocorrelation will be *inconclusive* with a Durbin-Watson test statistic of:

(a) 0.90
(b) 1.25
(c) 1.68
(d) 2.50
(e) 2.75

Answer: (d)
Difficulty level: Easy to Medium

14.11.7: Given n = 20, one independent variable, a null hypothesis of no positive autocorrelation, and α = .05, the null hypothesis will be accepted with a Durbin-Watson test statistic of:

(a) 0.80
(b) 1.10
(c) 2.00
(d) 2.50
(e) 2.85

Answer: (d)
Difficulty level: Easy to Medium

Statistical Process Control

15.2.1: A process that is in statistical control if the random variables associated with the outputs from the process over time are generated from:

(a)   a normal probability distribution
(b)   a binomial probability distribution
(c)   the same probability distribution and are statistically independent of one another
(d)   a different probability distribution and are statistically dependent
(e)   none of these

Answer: (c)
Difficulty level: Easy to Medium

15.2.2: A process that generates outputs over time with distributions that have the same mean, variance, and correlations at time lag k is known:

(a)   as a stationary process
(b)   as a nonstationary process
(c)   to be normally distributed
(d)   to follow a binomial distribution
(e)   a process that is not in control

Answer: (a)
Difficulty level: Easy to Medium

15.2.3: Which of the following factors may assignable causes be attributable to?

I.   differences in raw materials
II.   differences in employees
III. differences in machines

(a)   I only
(b)   II only
(c)   III only
(d)   I, II, and III
(e)   neither I nor II nor III

Answer: (d)
Difficulty level: Easy

15.2.4: Investing in new machinery or personnel development training systems are examples of attempts to reduce:

(a) productivity
(b) assignable or special cause variation
(c) common cause variation
(d) quality
(e) none of these

Answer: (c)
Difficulty level: Easy to Medium

15.2.5: Normal or natural variations in process outputs that are due purely to random or chance factors is an example of:

(a) productivity
(b) assignable or special cause variation
(c) common cause variation
(d) quality
(e) none of these

Answer: (b)
Difficulty level:

15.2.6: Distributions that are in control:

(a) cannot be improved
(b) are known to be normally distributed
(c) might be improved by reducing common cause variation
(d) might be improved by over-adjusting a stable process
(e) might be improved by reducing assignable cause variation

Answer: (c)
Difficulty level: Medium

15.2.7: According to the text, discovery and removal of assignable causes of instability in production processes are the responsibilities of _____.

(a) those people who are directly involved with the activity that generates the output distribution
(b) management
(c) all workers
(d) an impartial observer or mediator
(e) none of these

Answer: (a)
Difficulty level: Easy to Medium

15.2.8: According to the text, discovery and removal of common causes of instability in production processes are generally the responsibilities of _____.

(a)   those people who are directly involved with the activity that generates the output distribution
(b)   management
(c)   all workers
(d)   an impartial observer or mediator
(e)   none of these

Answer: (b)
Difficulty level: Easy to Medium

15.2.9: Which of the following statements are true?

I.   Statistical process control is continuously monitoring, isolating, and removing assignable causes of variation from processes to improve quality and productivity.

II.   Control charts are the fundamental method that is used for purposes of statistical process control.

III. A process that is out of control may have an assignable or special cause that produces a change in the distribution.

(a)   Only statement I is true
(b)   Only statements I and II are true
(c)   Only statements I and III are true
(d)   Statements I, II, and III are all true
(e)   None of the statements are true

Answer: (d)
Difficulty level: Easy to Medium

15.3.1: Control charts are used to:

(a)   monitor a process
(b)   warn that corrective action should be taken to remove a special cause of variation
(c)   indicate that a process should be changed to achieve a more favorable process or to reduce process variability
(d)   indicate when a process is in control
(e)   all of these

Answer: (e)
Difficulty level: Easy

15.3.2: A control chart that is used when the output of a
production process is measured in terms of the proportion of
defective units is:

(a)   an $\overline{X}$ chart
(b)   an I chart
(c)   a p chart
(d)   an R chart
(e)   none of these

Answer: (c)
Difficulty level: Easy to Medium

15.3.3: Which of the following is *not* an example of a
control chart that is used to monitor production processes?

(a) $\overline{X}$ charts
(b)   R charts
(c)   p charts
(d)   I charts
(e)   All of these are control charts that are commonly used
      to monitor production processes

Answer: (e)
Difficulty level: Easy

*Questions 15.3.4 - 15.3.5 refer to the following table that
summarizes subgroup sample means, sample variances, and
sample ranges (of each hour) for electrical charges of
memory chips produced by a manufacturer of such chips.*

A company manufactures memory chips for computers in an
ongoing process. The company randomly selected n = 4 memory
chips hourly for k = 12 consecutive hours, and the
measurements for the electrical charges (in microvolts) for
a conducting zone are summarized below.

			Subgroup (hour)			
Item	1	2	3	4	5	6
Mean	10.15	10.00	9.90	9.47	10.03	9.93
Variance	0.32	0.77	0.33	2.09	0.18	0.07
Range	1.20	2.10	1.30	3.50	1.00	0.60

Item	7	8	9	10	11	12
Mean	10.47	10.05	9.33	10.03	11.57	9.65
Variance	0.64	0.34	1.15	0.80	2.05	1.58
Range	1.60	1.10	2.00	2.10	3.00	2.80

15.3.4: The lower and upper control limits for an $\overline{X}$ chart are:

(a)   7.05 to 13.05
(b)   8.69 to 11.41
(c)   7.29 to 12.81
(d)   9.59 to 10.51
(e)   -10.05 to 10.05

Answer: (b)
Difficulty level: Easy to Medium

15.3.5: The lower and upper control limits for an R chart are:

(a)   -1.86 to 1.86
(b)   -1.86 to 0.00
(c)   0.00 to 1.86
(d)   0.00 to 4.24
(e)   0.00 to 2.28

Answer: (d)
Difficulty level: Easy to Medium

Questions 15.3.6 - 15.3.7 refer to the following table that summarizes the sample sizes, number of nonconforming chips, and fraction of nonconforming chips (of each hour) for memory chips produced by a manufacturer of such chips.

A company manufactures memory chips for computers in an ongoing process. The company randomly selected n = 4 memory chips hourly for k = 12 consecutive hours, and the measurements for the electrical charges (in microvolts) for a conducting zone are summarized below.

Item	Subgroup (hour)					
	1	2	3	4	5	6
Sample size	200	200	200	200	200	200
# Nonconforming	12	15	13	7	14	6
% Nonconforming	.060	.075	.065	.035	.070	.030

Item	7	8	9	10	11	12
Sample size	200	200	200	200	200	200
# Nonconforming	13	9	12	13	12	7
% Nonconforming	.065	.045	.060	.065	.060	.035

15.3.6: The lower and upper control limits for a p chart are:

(a)   -.055 to .055
(b)   -.0445 to .1545
(c)   -.048 to .048
(d)   .007 to .104
(e)   none of these

Answer: (d)
Difficulty level: Easy to Medium

15.3.7: Based on the above sample information, we should conclude that the fraction of nonconforming items produced by the process:

(a)   appears to be changing in a predictable pattern
(b)   does not appear to be changing in any predictable pattern
(c)   might be improved by reducing assignable cause variation
(d)   might be improved by reducing common cause variation
(e)   is out of control

Answer: (b)
Difficulty level: Medium

15.3.8: A production process is considered in control if at most 5% of the items being produced are defective. Samples of size 300 are used for the inspection process. The standard error of the proportion of defective items is:

(a)   .05
(b)   .0002
(c)   .0126
(d)   .0377
(e)   11.937

Answer: (c)
Difficulty level: Easy to Medium

15.3.9: A production process is considered in control if at most 5% of the items being produced are defective. Samples of size 300 are used for the inspection process. The lower and upper control limits for the p chart are:

(a)   .0123 and .0877
(b)   .04998 and .0502
(c)   .0374 and .0626
(d)   .01 and .09
(e)   not able to be determined with the information given

Answer: (a)
Difficulty level: Easy to Medium

15.3.10: A production process that is in control has a mean $\mu = 80$ and a standard deviation $\sigma = 10$. If five samples of size 25 yield sample means of 81, 84, 75, 83, and 79, is the process considered to be still in control?

(a)   Yes, since all of the sample ranges lie between the
       lower and upper limits of the R chart
(b)   Yes, since all of the sample means lie between the
       lower and upper limits of the X bar chart
(c)   No, since more than one of the sample means lies
       outside of the lower and upper limits of the X bar
       chart
(d)   We cannot say for sure because the fractions of
       nonconforming items are not given
(e)   We cannot say for sure because the sample variances are
       not given

Answer: (b)
Difficulty level: Medium

15.3.11: A production process that is in control has a mean of $\mu = 80$ and a standard deviation of $\sigma = 15$. The lower and upper control limits for an X bar chart based on samples of size 25 are:

(a)   71.0 and 89.0
(b)   78.2 and 81.8
(c)   77.68 and 82.32
(d)   35.0 and 125.0
(e)   not able to be determined with the information given

Answer: (a)
Difficulty level: Easy to Medium

15.3.12: An airline is interested in examining check-in times at its check-in counter. It randomly selects n = 4 passengers daily for k = 7 consecutive days for its 4:00 p.m. flight from an international airport to Los Angeles. The subgroup means, sample variances, and subgroup ranges were computed and are provided in the table below.

| | | | Subgroup (hour) | | | | |
Item	1	2	3	4	5	6	7
Mean	8.500	11.200	8.525	11.300	9.975	10.925	23.050
Variance	3.113	0.647	3.463	17.380	5.436	1.976	18.577
Range	3.50	1.80	4.30	9.70	4.90	2.80	9.00

a.   What are the lower and upper control limits for the $\overline{X}$ chart?
b.   Is the process considered to be in control?
c.   What are the lower and upper control limits for the R chart?
d.   Does the process variability appear to be in control?
e.   What are the values for the process capability indexes C_p and C_{pk} ?
f.   Is the check-in process capable of providing service that conforms to specifications for all of the customers?

Answer: a.   8.176 to 15.674
          b.   No. The seventh subgroup mean lies above the upper control limit
          c.   0.00 to 11.736
          d.   Yes, since all of the sample ranges lie between 0.00 and 11.736
          e. C_p = .62, C_{pk} = .61
          f.   No
Difficulty level: Easy to Medium

15.4.1: _____ uses data taken from a process that is in control to determine how well the process meets specifications for a product or service.

(a) Quality improvement
(b) Process capability analysis
(c) Control chart analysis
(d) Process control
(e) None of these

Answer: (b)
Difficulty level: Easy

15.4.2: The process capability index C_p is limited because:

(a)   it does not account for non-normal distributions
(b)   it does not account for the mean of the process
(c)   it does not account for the variance of the process
(d)   it can only take on positive values
(e)   all of these

Answer: (b)
Difficulty level: Medium

15.4.3: Engineers at Digital Industries, Inc., design memory chips to sort information accurately in a digital format. The memory chips will do so as long as the electrical charge for a conducting zone is between 7 and 13 microvolts. Measurements for n = 4 memory chips selected randomly by hour for k = 12 consecutive hours were taken. The subgroup mean and sample variances were recorded and are summarized in the table below.

Item	\multicolumn{6}{c}{Subgroup (hour)}
	1	2	3	4	5	6
Mean	10.15	10.00	9.90	9.47	10.03	9.93
Variance	0.32	0.77	0.33	2.09	0.18	0.07

Item	7	8	9	10	11	12
Mean	10.47	10.05	9.33	10.03	11.57	9.65
Variance	0.64	0.34	1.15	0.80	2.05	1.58

a.   What is the estimated standard deviation of the sampling distribution of the subgroup means?
b.   What is the estimated standard deviation of the output distribution?
c.   What are the standardized Z values for the lower and upper specification limits?
d.   What is the estimated proportion of Digital's memory chips that conform to specifications?
e.   What are the values of the process capability indexes C_p and C_{pk} ?

Answer: a.   .45
          b.   .92
          c.   -3.71 and 3.49
          d.   .4995 + .4993 = .9988
          e. C_p = 1.09, C_{pk} = 1.07
Difficulty level: Easy to Medium

15.5.1: A financial analyst at a large bank wanted to examine the changes over time of short term interest rates. The analyst decided to do so by examining the yields on 90-day treasury bills. She collected monthly yield data for a period of two years. The results are presented in order (left to right from first row to last row) in the table below. Do the yields appear to be generated from a stable process?

7.76	8.22	8.57	8.00	7.56	7.01	7.05	7.18
7.08	7.17	7.20	7.07	7.04	7.03	6.59	5.49
6.12	6.21	5.84	5.57	5.19	5.18	5.35	5.49

Answer: No. The time series plot of yields for 90-day treasury bills has a downward trend and is hence not a stable process
Difficulty level: Easy to Medium

15.5.2: A large lighting and fixture company randomly selected n = 50 lamps over 50 days and the measurements for the lifetimes in hours of continuous operation and the moving ranges were measured, yielding a mean lifetime of the individual data is equal to 24.114, and a sample mean for the moving range values of 2.739, the upper control limit for the I-chart is:

(a)   16.828
(b)   31.400
(c)   24.114
(d)   2.739
(e)   7.286

Answer: (b)
Difficulty level: Easy

15.5.3: A large lighting and fixture company randomly selected n = 50 lamps over 50 days and the measurements for the diameters of individual lamps were measured, yielding a mean diameter of the individual data is equal to 2.001, and a sample mean for the moving range values of 0.0147, the lower control limit for the I-chart is:

(a)   1.962
(b)   2.040
(c)   0.0391
(d)   2.001
(e)   0.0147

Answer: (b)
Difficulty level: Easy

15.5.4: A large company produces valves for automobile engines by using a hot-metal-forging manufacturing process. To examine the diameters of the valves prior to machining and grinding, an engineer from the company randomly selected n = 4 valves hourly for k = 8 consecutive hours during a production shift. The diameters were recorded in centimeters. The subgroup means, sample variances, and subgroup ranges are summarized in the table below. Answer the following.

Item	Subgroup (hour)							
	1	2	3	4	5	6	7	8
Mean	4.975	5.125	4.90	5.075	5.050	5.025	5.850	5.90
Variance	.0025	.0492	.100	.0225	.0367	.2025	.1967	.0067
Range	.10	.50	.70	.30	.40	1.00	1.00	.20

a. What are the lower and upper control limits for the $\bar{X}$ chart?
b. Do any of the sample means fall outside of the control limits? Is the process considered to be in control?
c. If the process is not in control, then what assignable cause is likely to be influencing the process?
d. What are the lower and upper control limits for the R chart?
e. Does the process variability appear to be in control?
f. What proportion of outputs conform to specifications?
g. What are the values for the process capability indexes C_p and C_{pk} ?
h. Is the process capable of providing outputs that conform to the specifications for all of its products?

Answer: a.   4.85 to 5.62
          b.   Yes. No, the process is not in control.
          c.   Employees may be tiring at the end of shifts.
          d.   .00 and 1.20
          e.   Yes.
          f.   .9681
          g. C_p = .72, C_{pk} = .68
          h.   No
Difficulty level: Easy to Medium

15.5.5: A local company produces drive shafts for small
electrical motors by using a numerically controlled milling
machine.   To monitor the manufacturing process, the company
randomly selected n = 4 drive shafts daily for k = 10
consecutive working days.   The lengths of the drive shafts
were measured in centimeters.   The subgroup means, sample
variances, and subgroup ranges are summarized in the table
below.

Item	Subgroup (hour)				
	1	2	3	4	5
Mean	9.45	9.875	10.000	9.775	9.600
Variance	.310	.12917	1.320	.4025	.14667
Range	1.30	0.80	2.40	1.50	0.80

Item	6	7	8	9	10
Mean	9.975	9.825	10.300	9.475	10.375
Variance	.4625	.6825	.7600	.58917	.80917
Range	1.60	1.80	1.80	1.70	2.00

a.   What are the lower and upper control limits for the $\overline{X}$
      chart?
b.   Do any of the sample means fall outside of the control
      limits?   Is the process considered to be in control?
c.   If the process is not in control, then what assignable
      cause is likely to be influencing the process?
d.   What are the lower and upper control limits for the R
      chart?
e.   Does the process variability appear to be in control?

Answer: a.   8.72 and 11.01
           b.   No.   Yes, the process is in control.
           c.   None
           d.   .00 and 3.58
           e.   Yes.
Difficulty level: Easy to Medium

Decision Analysis

16.1.1: Statistical decision making *requires* knowledge of:

I.   the economic consequences of alternative actions
II. the set of actions available to the decision maker
III. the probabilities that various economic consequences will
     be realized
IV. the results of an action are expressed in terms of
     monetary values

(a) IV only
(b) I and II only
(c) I and III only
(d) I, II, and III only
(e) I, II, III, and IV

Answer: (d)
Difficulty level: Easy to Medium

16.1.2: One action dominates another when:

(a) most of its payoffs are better than the other's
     regardless of what event occurs
(b) all of its payoffs are equal to or better than the
     other's regardless of what event occurs
(c) all of its payoffs are less than or equal to or better
     than the other's regardless of what event occurs
(d) it results in a larger payoff for an event that is very
     likely to occur
(e) none of these

Answer: (b)
Difficulty level: Easy

16.1.3: A graphic presentation of the expected gain from the
various options available to the decision maker is called:

(a) a decision tree
(b) a payoff table
(c) the expected monetary value
(d) an opportunity loss table
(e) the expected value of perfect information

Answer: (b)
Difficulty level: Easy

16.2.1: The probabilities entered into a payoff table are called _____ probabilities.

(a)   conditional
(b)   prior
(c)   Bayesian
(d)   revised
(e)   none of these

Answer: (b)
Difficulty level: Easy

16.3.1: Which of the following statements are correct?

I.   Entering the probabilities of each event into a payoff table directly indicates which action should be taken

II.   The expected value of perfect information is the amount by which the expected payoff under certainty differs from the largest expected monetary value

III.   If a decision analysis problem is *sequential*, in that it calls for a sequence of decisions to be made, then the problem can adequately be represented with a payoff table

(a)   I only
(b)   II only
(c)   III only
(d)   II and III only
(e)   none of these statements are correct

Answer: (b)
Difficulty level: Easy to Medium

16.3.2: The action with the highest weighted payoff is the one having:

(a)   the smallest expected monetary value
(b)   the largest expected monetary value
(c)   the highest probability of occurrence
(d)   the lowest probability of occurrence
(e)   none of these

Answer: (b)
Difficulty level: Easy to Medium

16.4.1: A tabular presentation of the expected gain from the various options available to the decision maker is called:

(a) a decision tree
(b) a payoff table
(c) the expected monetary value
(d) an opportunity loss table
(e) the expected value of perfect information

Answer: (a)
Difficulty level: Easy

16.4.2: The principle of *backward induction* indicates that the best act can be selected at a decision node only if:

(a) the action which minimizes the expected opportunity loss is chosen
(b) the action which maximizes the expected monetary value is chosen
(c) the best act at decision nodes to the right have already been selected
(d) the best act at decision nodes to the left have already been selected
(e) none of these

Answer: (c)
Difficulty level: Easy to Medium

16.4.3: Sensitivity analysis specifically relates to the:

(a) decision tree decision nodes
(b) configuration of the opportunity loss table
(c) setup of the payoff table
(d) setup of the decision tree
(e) changing of probabilities

Answer: (e)
Difficulty level: Easy to Medium

16.5.1: The expected opportunity loss, EOL, is equal to:

(a) the value of perfect information
(b) the payoff under uncertainty
(c) the expected cost of uncertainty
(d) the value of utility
(e) the expected monetary value

Answer: (c)
Difficulty level: Easy

16.6.1: Which relationship is important because it gives an upper limit on how much a decision maker should be willing to spend to eliminate the uncertainty facing them?

(a) EVPI = Expected monetary value
(b) EVPI = Expected payoff under certainty
(c) EPUC = Expected monetary value
(d) EOL = Expected cost of uncertainty
(e) EOL = Expected payoff under certainty

Answer: (d)
Difficulty level: Medium

16.6.2: When the expected value of perfect information equals zero, the expected payoff under certainty must be:

(a) equal to the expected monetary value
(b) smaller than the expected monetary value
(c) equal to the expected opportunity loss
(d) smaller than the expected opportunity loss
(e) greater than the expected monetary value

Answer: (a)
Difficulty level: Easy

16.7.1: Which of the following decision criteria ignore the probabilities of each event occurring?

(a) maximax
(b) minimax
(c) maximum likelihood
(d) expected monetary value
(e) more than one of these

Answer: (a)
Difficulty level: Easy to Medium

16.7.2: Where an organization or person is in financial difficulty and cannot afford a large loss, the appropriate decision criterion is the:

(a) Bayesian criterion
(b) equal likelihood criterion
(c) maximum likelihood criterion
(d) maximin criterion
(e) maximax criterion

Answer: (d)
Difficulty level: Easy

16.7.3: When an event has an extremely high probability, or time constraints dictate that the event with the highest probability only be analyzed, the decision maker should use:

(a)   a Bayesian criterion
(b)   an expected monetary value criterion
(c)   a maximum likelihood criterion
(d)   a maximin criterion
(e)   a maximax criterion

Answer: (c)
Difficulty level: Easy to Medium

16.7.4: Except for ties, the decision maker would be well off if he used only the criterion of:

(a)   equal likelihood
(b)   expected monetary value
(c)   maximum likelihood
(d)   maximin
(e)   maximax

Answer: (b)
Difficulty level: Easy to Medium

16.8.1: The assumption of risk neutrality is likely to be an accurate description of a business decision maker's behavior:

(a)   if potential losses are a large fraction of firm assets
(b)   if potential losses are a small fraction of firm assets
(c)   if potential gains are a large fraction of firm assets
(d)   if potential gains are a small fraction of firm assets
(e)   if potential losses equal potential gains

Answer: (b)
Difficulty level: Medium

16.8.2: When the payoffs involved are very large relative to the overall wealth, assets, or budget of the decision maker:

(a)   expected monetary value is a good decision-making criterion
(b)   a utility criterion should be considered
(c)   a utility criterion should not be considered
(d)   expected monetary value equals expected opportunity loss
(e)   none of these

Answer: (b)
Difficulty level: Easy to Medium

Use the accompanying table to answer questions 16.9.1 - 16.9.5

An owner of a business faces the following situation. She must decide between selling now (A_1) or holding her business and waiting for one year (A_2). Assume that her situation is described by the table below, which reflects payoffs that are discounted to present value.

EVENT	Probability	OWNER'S ACTION A_1	A_2
E_1, Contract given	.20	80	100
E_2, Contract not given	.80	80	70

16.9.1: The expected monetary value of the business owner's action to sell now (A_1) is:

(a) $4,000
(b) $76,000
(c) $80,000
(d) $8,000
(e) $84,000

Answer: (c)
Difficulty level: Easy

16.9.2: The expected opportunity loss of the business owner's action to wait a year (A_2) is:

(a) $4,000
(b) $76,000
(c) $80,000
(d) $8,000
(e) $84,000

Answer: (d)
Difficulty level: Easy

16.9.3: Using expected monetary value, the best decision is:

(a) Sell the business now
(b) Wait a year
(c) Make sure you get the contract, then wait a year
(d) Sell the business only if you do not get the contract
(e) None of these

Answer: (a)
Difficulty level: Easy

16.9.4: Using expected opportunity loss, the best decision is:

(a)   Sell the business now
(b)   Wait a year
(c)   Make sure you get the contract, then wait a year
(d)   Sell the business only if you do not get the contract
(e)   None of these

Answer: (a)
Difficulty level: Easy

16.9.5: The dollar value the owner of the business would be willing to spend to gain more information about her chances of winning the contract is equal to:

(a) $4,000
(b) $76,000
(c) $80,000
(d) $8,000
(e) $84,000

Answer: (a)
Difficulty level: Easy to Medium

16.9.6: The manager of a large concrete company has an opportunity to buy up to three carloads of cement. Another company has told the manager that they are willing to buy some of the cement from him, but has not told him how many carloads it will take. Any cement that is not bought by the other company can be sold to a broker. Assume that the manager's situation is described by the payoff table below.

OTHER COMPANY'S ORDER	MANAGER'S ACTION			
	Buy 0	Buy 1	Buy 2	Buy 3
Order 0	0	-600	-1200	-1800
Order 1	0	1800	1200	600
Order 2	0	1800	3600	3000
Order 3	0	1800	3600	5400

The payoff table above is one:

(a)   with no dominant actions or strategies
(b)   with a single dominant action or strategy
(c)   with two dominant actions or strategies
(d)   with three dominant actions or strategies
(e)   none of these

Answer: (a)
Difficulty level: Easy to Medium

Use the accompanying table to answer questions 16.9.7 - 16.9.9

A large semiconductor company must decide whether to market a new product. The company's profit or loss on the product depends on the percentage of the market it can capture. The break-even market share is 10%. The company's subjective beliefs about its chances at various possible market shares are reflected in the payoff table below.

| EVENT | Probability | COMPANY'S ACTION | |
		Market	Do not market
7% Market Share	.050	-110	0
8% Market Share	.100	-80	0
9% Market Share	.120	-50	0
10% Market Share	.180	0	0
11% Market Share	.250	40	0
12% Market Share	.200	80	0
13% Market Share	.100	120	0

16.9.7: Using expected monetary value, the best decision is:

(a) is to market the new product
(b) is to not market the new product
(c) is to get a 12% market share, then market the new product
(d) is to get a 13% market share, then market the new product
(e) none of these

Answer: (a)
Difficulty level: Easy to Medium

16.9.8: The expected payoff under certainty is:

(a) EPUC = $0
(b) EPUC = $18,500
(c) EPUC = $38,000
(d) EPUC = $19,500
(e) None of these

Answer: (c)
Difficulty level: Easy to Medium

16.9.9: The expected value of perfect information is:

(a)   EVPI = $0
(b)   EVPI = $18,500
(c)   EVPI = $38,000
(d)   EVPI = $19,500
(e)   None of these

Answer: (d)
Difficulty level: Easy to Medium

Use the table below to answer questions 16.9.10 - 16.9.14

A company is considering a large construction project. The table below presents the *costs* (in thousands of dollars) for the construction project where there are four alternative methods of construction (A_1, A_2, A_3, and A_4) and three types of weather conditions that might be encountered (E_1, E_2, and E_3).

		COMPANY'S ACTION			
EVENT	Probability	$\underline{A_1}$	$\underline{A_2}$	$\underline{A_3}$	$\underline{A_4}$
E_1	.20	20	20	26	30
E_2	.30	25	30	35	30
E_3	.50	20	18	10	22

16.9.10: The method of construction available to the company with the highest expected monetary value is:

(a) A_1, with an expected monetary value of $21,500
(b) A_2, with an expected monetary value of $22,000
(c) A_3, with an expected monetary value of $20,700
(d) A_4, with an expected monetary value of $26,000
(e) A_3, with an expected monetary value of $26,000

Answer: (d)
Difficulty level: Easy

16.9.11: Are there any dominated actions or strategies?

(a)   Yes, only A_1 is dominated by A_4
(b)   Yes, only A_2 is dominated by A_4
(c)   Yes, both A_1 and A_2 are dominated by A_4
(d)   Yes, only A_4 is dominated by A_1, A_2, and A_3
(e)   No, there are no dominated strategies

Answer: (c)
Difficulty level: Medium

16.9.12: Using expected monetary value, the company should:

(a)   choose the A_1 method of construction
(b)   choose the A_2 method of construction
(c)   choose the A_3 method of construction
(d)   choose the A_4 method of construction
(e)   choose any of the four methods of construction

Answer: (d)
Difficulty level: Easy to Medium

16.9.13: The expected payoff under certainty is:

(a)   EPUC = $22,000
(b)   EPUC = $27,500
(c)   EPUC = $26,000
(d)   EPUC = $1,500
(e)   None of these

Answer: (b)
Difficulty level: Easy to Medium

16.9.14: The expected value of perfect information is:

(a)   EVPI = $22,000
(b)   EVPI = $27,500
(c)   EVPI = $26,000
(d)   EVPI = $1,500
(e)   None of these

Answer: (d)
Difficulty level: Easy to Medium

Use the table below to answer questions 16.9.15 - 16.9.19

The following represents a payoff table (in thousands of dollars) for a large computer leasing company's three leasing alternative sizes (large, medium, and small) and the two states of nature (high acceptance, E_1, and low acceptance, E_2).

EVENT	Probability	ACTION (Leasing Size)		
		Large	Medium	Small
High acceptance (E_1)	.30	200	150	100
Low acceptance (E_2)	.70	-20	20	60

16.9.15: Using the maximax criterion, the company should:

(a) Choose the small leasing size
(b) Choose the medium leasing size
(c) Choose the large leasing size
(d) Not lease at all
(e) None of these

Answer: (c)
Difficulty level: Easy

16.9.16: Using expected monetary value, the best decision is:

(a) the small leasing size, since its EMV is the largest
(b) the medium leasing size, since its EMV is the largest
(c) the large leasing size, since its EMV is the largest
(d) the large leasing size, since its EMV is the smallest
(e) the medium leasing size, since its EMV is the smallest

Answer: (a)
Difficulty level: Easy

16.9.17: Using expected opportunity loss, the best decision:

(a) the small leasing size, since its EOL is the largest
(b) the medium leasing size, since its EOL is the largest
(c) the large leasing size, since its EOL is the smallest
(d) the small leasing size, since its EOL is the smallest
(e) the medium leasing size, since its EOL is the smallest

Answer: (d)
Difficulty level: Easy

16.9.18: The expected payoff under certainty is:

(a) EPUC = $72,000
(b) EPUC = $30,000
(c) EPUC = $13,000
(d) EPUC = $59,000
(e) EPUC = $102,000

Answer: (e)
Difficulty level: Easy to Medium

16.9.19: The expected value of perfect information is:

(a)   EVPI = $72,000
(b)   EVPI = $30,000
(c)   EVPI = $13,000
(d)   EVPI = $59,000
(e)   EVPI = $102,000

Answer: (b)
Difficulty level: Easy to Medium

Use the table below to answer questions 16.9.20 - 16.9.24

The following represents a payoff table (in thousands of dollars) for a large computer leasing company. The table summarizes the company's three leasing alternative sizes (large, medium, and small) and the two states of nature (high acceptance and low acceptance).   In accordance with the maximax criterion, the company should:

		ACTION (Leasing Size)		
EVENT	**Probability**	Large	Medium	Small
High acceptance (E_1)	.60	200	150	100
Low acceptance (E_2)	.40	-20	20	60

16.9.20: Using the pessimist's criterion, the company should choose the _____ leasing size because _____.

(a)   small; its EMV is the smallest
(b)   small; its EOL is the largest
(c)   medium; its EOL is in the middle
(d)   large; its EMV is the smallest
(e)   large; its EOL is the smallest

Answer: (e)
Difficulty level: Easy to Medium

16.9.21: Using expected monetary value, the best decision is:

(a)   small; its EMV is the smallest
(b)   small; its EMV is the largest
(c)   medium; its EMV is in the middle
(d)   large; its EMV is the smallest
(e)   large; its EMV is the largest

Answer: (e)
Difficulty level: Easy to Medium

16.9.22: Using expected opportunity loss, the best decision:

(a)   small; its EOL is the smallest
(b)   small; its EOL is the largest
(c)   medium; its EOL is in the middle
(d)   large; its EOL is the smallest
(e)   large; its EOL is the largest

Answer: (d)
Difficulty level: Easy to Medium

16.9.23: The expected payoff under certainty is:

(a)   EPUC = $84,000
(b)   EPUC = $98,000
(c)   EPUC = $112,000
(d)   EPUC = $144,000
(e)   EPUC = $32,000

Answer: (d)
Difficulty level: Easy to Medium

16.9.24: The expected value of perfect information is:

(a)   EVPI = $84,000
(b)   EVPI = $98,000
(c)   EVPI = $112,000
(d)   EVPI = $144,000
(e)   EVPI = $32,000

Answer: (e)
Difficulty level: Easy to Medium

Use the table below to answer questions 16.9.25 - 16.9.30

International Container, Inc., must decide where it will locate a new plant to produce a new packing product. The plant is to be constructed in London, Beijing, or New York (A_1, A_2, and A_3), and the major market for the product may be in Europe, the Far East, or the United States (E_1, E_2, and E_3). The profits (in millions of dollars) for the new product as presented in the payoff table below.

		ACTION (Plant Location)		
EVENT	Probability	London	Beijing	New York
Europe	.45	40	12	-5
Far East	.30	-10	50	18
United States	.25	28	-2	20

16.9.25: The expected monetary value of the company choosing
to locate its new packing plant in Beijing is:

(a)   19.9 million dollars
(b)   22.0 million dollars
(c)   40.0 million dollars
(d)   18.0 million dollars
(e)   20.1 million dollars

Answer: (a)
Difficulty level: Easy

16.9.26: Using expected monetary value, the best decision is:

(a)   to locate in London, since its EMV is the largest
(b)   to locate in Beijing, since its EMV is the largest
(c)   to locate in New York, since its EMV is the largest
(d)   to locate in London, since its EOL is the smallest
(e)   to locate in New York, since its EMV is the smallest

Answer: (a)
Difficulty level: Easy

16.9.27: The expected opportunity loss of the company choosing
to locate its new packing plant in London is:

(a)   19.9 million dollars
(b)   20.1 million dollars
(c)   22.0 million dollars
(d)   8.15 million dollars
(e)   none of these

Answer: (e)
Difficulty level: Easy to Medium

16.9.28: Using expected opportunity loss, the best decision:

(a)   to locate in London, since its EOL is the largest
(b)   to locate in Beijing, since its EOL is the smallest
(c)   to locate in New York, since its EOL is the largest
(d)   to locate in London, since its EOL is the smallest
(e)   to locate in New York, since its EOL is the smallest

Answer: (d)
Difficulty level: Easy to Medium

16.9.29: The expected payoff under certainty is:

(a) EPUC = 40.0 million dollars
(b) EPUC = 18.0 million dollars
(c) EPUC = 20.1 million dollars
(d) EPUC = 22.0 million dollars
(e) None of these

Answer: (a)
Difficulty level: Easy to Medium

16.9.30: The expected value of perfect information is:

(a) EVPI = 40.0 million dollars
(b) EVPI = 18.0 million dollars
(c) EVPI = 20.1 million dollars
(d) EVPI = 22.0 million dollars
(e) None of these

Answer: (b)
Difficulty level: Easy to Medium

Use the table below to answer questions 16.9.31 - 16.9.36

A real estate investor currently has an option to purchase some property. The value of the property depends on whether a City Planning Commission rezones it to the classification that the real estate investor desires. The investor has three actions: do nothing, renew, and buy outright. The planning commission can make three decisions: rezone, deny, and table. The table below summarizes payoffs in thousands of dollars.

		ACTION (Investor decision)		
EVENT	Probability	Do nothing	Renew	Buy outright
E_1, Rezone	.15	0	85	100
E_2, Deny	.30	0	-10	-15
E_3, Table	.55	0	5	20

16.9.31: The expected monetary value of the real estate investor deciding to "buy outright" is:

(a) $0
(b) $12,500
(c) $21,500
(d) $26,000
(e) $4,500

Answer: (c)
Difficulty level: Easy

16.9.32: Using expected monetary value, the best decision for the real estate investor to make is:

(a)   Do nothing since its EMV is the largest
(b)   Do nothing since its EMV is the smallest
(c)   Renew since its EMV is in the middle
(d)   Buy outright since its EMV is the largest
(e)   Buy outright since its EMV is the smallest

Answer: (d)
Difficulty level: Easy

16.9.33: The expected opportunity loss of the real estate investor deciding to "buy outright" is:

(a) $0
(b) $12,500
(c) $21,500
(d) $26,000
(e) $4,500

Answer: (e)
Difficulty level: Easy

16.9.34: Using expected opportunity loss, the best decision for the real estate investor is:

(a)   Do nothing since its EOL is the largest
(b)   Do nothing since its EOL is the smallest
(c)   Renew since its EOL is in the middle
(d)   Buy outright since its EOL is the largest
(e)   Buy outright since its EOL is the smallest

Answer: (e)
Difficulty level: Easy to Medium

16.9.35: The expected payoff under certainty is:

(a) $0
(b) $12,500
(c) $21,500
(d) $26,000
(e) $4,500

Answer: (d)
Difficulty level: Easy to Medium

16.9.36: The expected value of perfect information is:

(a) $0
(b) $12,500
(c) $21,500
(d) $26,000
(e) $4,500

Answer: (e)
Difficulty level: Easy to Medium

Use the table below to answer questions 16.9.37 - 16.9.42

The accompanying payoff table presents the profits for a
company manufacturing operation (in thousands of dollars),
where there are three alternative machines that can be
selected (A_1, A_2, and A_3) to produce a product, and the demand
for the product might by high, medium, or low (E_1, E_2, and E_3).

		ACTION	(Machine	Selected)
EVENT	Probability	A_1	A_2	A_3
High demand (E_1)	.20	7	28	-9
Medium demand (E_2)	.35	10	-20	1
Low demand (E_3)	.45	-8	13	15

16.9.37: The expected monetary value of the company choosing
machine 2 to produce a product is:

(a) $14,550
(b) $11,400
(c) $10,550
(d) $4,450
(e) $5,300

Answer: (d)
Difficulty level: Easy

16.9.38: Which machine should the company select on the basis
of expected monetary value?

(a)   Machine 1
(b)   Machine 2
(c)   Machine 3
(d)   Either machine 1 or machine 2
(e)   Either machine 2 or machine 3

Answer: (c)
Difficulty level: Easy

16.9.39: The value of the expected opportunity loss of the company choosing machine 1 to produce a product is:

(a) $14,550
(b) $11,400
(c) $10,550
(d) $4,450
(e) $5,300

Answer: (a)
Difficulty level: Easy to Medium

16.9.40: Which machine should the company select on the basis of expected opportunity loss?

(a) Machine 1
(b) Machine 2
(c) Machine 3
(d) Either machine 1 or machine 2
(e) Either machine 2 or machine 3

Answer: (c)
Difficulty level: Easy

16.9.41: The expected payoff under certainty is:

(a) $15,850
(b) $11,400
(c) $10,550
(d) $4,450
(e) $5,300

Answer: (a)
Difficulty level: Easy to Medium

16.9.42: The expected value of perfect information is:

(a) $15,850
(b) $11,400
(c) $10,550
(d) $4,450
(e) $5,300

Answer: (c)
Difficulty level: Easy

Decision Analysis with Sample Information

17.1.1: Which of the following statements are true?

I. Decision analysis with sample information concentrates on the uncertainty of the events that might take place and how this uncertainty can be measured and expressed.

II. Decision makers often use the results of a survey, sample, or experiment to obtain better estimates of subjectively determined probabilities.

III. Obtaining a probability curve is more burdensome than finding the subjective probabilities for all of the events that might occur

(a) I only
(b) II only
(c) I and II only
(d) I, II, and III
(e) None of these statements are true

Answer: (c)
Difficulty level: Easy to Medium

17.1.2: Decision makers have to resort to using subjective probability when:

(a) Facing a problem one has never faced before.
(b) There is no historical information to guide their assessment.
(c) Building an experimental rocket never produced before.
(d) All of these.
(e) They have little or no idea of how likely it is for each of the possible events to occur

Answer: (d)
Difficulty level: Easy

17.2.1: A construction company has been told that it will be awarded a contract to build a large flood control system, contingent upon the state legislature's votes for a bill that approves of funds for the project. The company's president believes that there is a .60 chance that the bill will pass and a .40 chance that it will not. Company managers who are unsure whether to begin substantive planning for the project decide to look at information published in a local newspaper. In the past, the newspaper the following accuracy record:

P(Predict passage | Pass)= .80, P(Predict Failure | Pass)= .20
P(Predict Passage | Fail)= .05, P(Predict Failure | Fail)= .95

Compute the following probabilities.

a.   P(Bill passes | Newspaper predicts passage).
b.   P(Bill fails | Newspaper predicts passage).
c.   P(Bill will pass).
d.   P(Bill will fail).

Answer: a.   .96
           b.   .04
           c.   .60
           d.   .40
Difficulty level: Medium

17.2.2: Historical records for the production output of a machine show that the proportion of defectives produced by the machine is as follows:

Proportion of Defectives	Percentage of Batches with Proportion p Defectives
.01	.70
.05	.20
.10	.10

A sample of n = 20 items has just been taken from a new batch of the machine's output, and one was found to be defective.

What is the revised probability that the true proportion of defectives is .01?

(a)   .5300
(b)   .3462
(c)   .1238
(d)   .0000
(e)   1.000

Answer: (a)
Difficulty level: Medium

17.2.3: Historical records for the production output of a machine show that the proportion of defectives produced by the machine is as follows:

Proportion of Defectives	Percentage of Batches with Proportion p Defectives
.01	.70
.05	.20
.10	.10

A sample of n = 20 items has just been taken from a new batch of the machine's output, and one was found to be defective.

What is the revised probability that the true proportion of defectives is .10?

(a)   .5300
(b)   .3462
(c)   .1238
(d)   .0000
(e)   1.000

Answer: (c)
Difficulty level: Medium

17.3.1: When deciding whether it is appropriate to revise ones prior probabilities by using a sample or experiment, one should usually obtain the sample information when:

(a)   the expected monetary value is high
(b)   the expected payoff under certainty us high
(c)   the expected value of perfect information is large and the cost of the experiment is relatively large
(d)   the expected value of perfect information is large and the cost of the experiment is relatively small
(e)   the expected value of perfect information is small and the cost of the experiment is relatively large

Answer: (d)
Difficulty level: Easy

17.3.2: A bank president is worried about actions the Federal Reserve may take next week. After the meeting of the Federal Reserve she feels that there is a 40% chance that credit policies will be tightened up, making it more difficult to loan money, and a 60% chance that credit policies will be loosened up, making it easier to loan money. To get more information concerning the chances that credit will be tighter or looser after next week, she contacted an economist, who predicted that the Federal Reserve will loosen credit policies. The economist's accuracy record is being correct 80% of the time that credit policies were loosened and incorrect 15% of the time where credit was tightened (and he predicted a loosening). Compute the following probabilities.

a.   What is the probability that the Federal Reserve loosens its credit policies next week <u>and</u> the economist predicted the loosening?
b.   What is the probability that the Federal Reserve tightens its credit policies next week <u>and</u> the economist predicted a loosening (of credit policies)?

Answer: a.   P(Fed loosens credit and economist predicts loosening) = (.80)(.60) = .48
          b.   P(Fed tightens credit and economist predicts loosening) = (.15)(.40) = .06
Difficulty level: Medium

17.4.1: Assume that three states of nature A, B, and C have prior probabilities of 0.60, 0.30, and 0.10, respectively. The conditional probabilities of observing experimental result E, given A, B, and C, are 0.20, 0.50, and 0.30, respectively.

If the experimental result E is observed, compute the following probabilities.

a. $P(A|E)$?
b. $P(B|E)$?
c. $P(C|E)$?
d.   P(A and E)?
e.   P(B and E)?
f.   P(C and E)?

Answer: a. $P(A|E)$ = .40
          b. $P(B|E)$ = .50
          c. $P(C|E)$ = .10
          d.   P(A and E) = .12
          e.   P(B and E) = .15
          f.   P(C and E) = .03
Difficulty level: Easy to Medium

17.4.2: Assume that three states of nature A, B, and C have prior probabilities of 0.60, 0.30, and 0.10, respectively. The conditional probabilities of observing experimental result E, given A, B, and C, are 0.25, 0.25, and 0.50, respectively.

If the experimental result E is *not* observed, compute the following probabilities (assuming that experimental result F is the complement of experimental result E).

a. P(A|F)?
b. P(B|F)?
c. P(C|F)?
d. P(A and F)?
e. P(B and F)?
f. P(C and F)?

Answer: a.   P(A|F) = .6207
           b.   P(B|F) = .3103
           c.   P(C|F) = .0690
           d.   P(A and F) = .45
           e.   P(B and F) = .225
           f.   P(C and F) = .05
Difficulty level: Medium

17.4.3: Assume that three states of nature A, B, and C have prior probabilities of 0.50, 0.20, and 0.30, respectively. The conditional probabilities of observing experimental result E, given A, B, and C, are 0.30, 0.60, and 0.80, respectively.

If the experimental result E is observed, compute the following probabilities.

a. P(A|E)?
b. P(B|E)?
c. P(C|E)?
d. P(A and E)?
e. P(B and E)?
f. P(C and E)?

Answer: a.   P(A|E) = .2941
           b.   P(B|E) = .2353
           c.   P(C|E) = .4706
           d.   P(A and E) = .15
           e.   P(B and E) = .12
           f.   P(C and E) = .24
Difficulty level: Easy to Medium

17.4.4: Assume that three states of nature A, B, and C have prior probabilities of 0.20, 0.10, and 0.70, respectively. The conditional probabilities of observing experimental result E, given A, B, and C, are 0.30, 0.30, and 0.50, respectively.

If the experimental result E is *not* observed, compute the following probabilities (assuming that experimental result F is the complement of experimental result E).

a. $P(A|F)$?
b. $P(B|F)$?
c. $P(C|F)$?
d. $P(A$ and $F)$?
e. $P(B$ and $F)$?
f. $P(C$ and $F)$?

Answer: a. $P(A|F) = .250$
          b. $P(B|F) = .125$
          c. $P(C|F) = .625$
          d. $P(A$ and $F) = .14$
          e. $P(B$ and $F) = .07$
          f. $P(C$ and $F) = .35$
Difficulty level: Medium

17.4.5: A mining company is attempting to decide whether or not to purchase a certain piece of land. The land costs $300,000. If there are commercial ore deposits on the land, the company can sell the property for $500,000. If no ore deposits exist, however, the property will sell for $200,000. The company believes that the odds are 50-50 that commercial ore deposits are present. Before purchasing the land, the property can be test cored at a cost of $20,000. The coring will indicate if conditions are favorable or unfavorable for core mining. If there are commercial deposits present, the probability of a favorable core report is 85%. If such ore deposits are not present, the probability of a favorable report is only 20%.

a.   What is the (posterior) probability of getting a favorable report? What is the (posterior) probability of getting an unfavorable report?
b.   What decision should the company make?
c.   If the company did not have the option to test the land, what would the expected value of perfect information be?

Answer: a.   P (Favorable) = .525, P (Unfavorable) = .475
          b.   Test the land; if the test is favorable, buy the land; if the test is unfavorable, do not buy the land.
          c. $50,000
Difficulty level: Easy to Medium

Use the tables below to answer questions 17.4.6 - 17.4.9.

The following represents a payoff table (in thousands of dollars) for a large computer leasing company that summarizes the payoffs and probabilities for the company's three leasing alternative sizes (large, medium, and small) and the two states of nature (high acceptance and low acceptance).

EVENT	Probability	**ACTION (Leasing size)**		
		Large	Medium	Small
High Acceptance (E_1)	.30	200	150	100
Low Acceptance (E_2)	.70	-20	20	60

EVENT	**Marketing Research Report**	
	Favorable (F)	Unfavorable (U)
High acceptance (E_1)	$P(F \mid E_1) = .80$	$P(U \mid E_1) = .20$
Low acceptance (E_2)	$P(F \mid E_2) = .10$	$P(U \mid E_2) = .90$

17.4.6: The probability that the market research report will be favorable; $P(F)$ is:

(a)   .80
(b)   .24
(c)   .31
(d)   .7742
(e)   .2258

Answer: (c)
Difficulty level: Medium

17.4.7: The probability that the observed event will be high acceptance, given a favorable market research report is:

(a)   .80
(b)   .24
(c)   .31
(d)   .7742
(e)   .2258

Answer: (d)
Difficulty level: Medium

17.4.8: The expected value of the market research information:

(a)   is $90,402
(b)   is $72,000
(c)   is $30,000
(d)   is $18,402
(e)   is $0

Answer: (d)
Difficulty level: Medium to Challenging

17.4.9: The company's optimal decision strategy is:

(a)   if the market research report is favorable, then lease
      the small system, if the market research report is not
      favorable, then lease the large system
(b)   if the market research report is favorable, then lease
      the large system, if the market research report is not
      favorable, then lease the small system
(c)   lease the medium
(d)   lease the large system
(e)   lease the small system

Answer: (b)
Difficulty level: Medium

Use the tables below to answer questions 17.4.10 - 17.4.12.

A manufacturing company has the following payoff table for a
make-or-buy decision:

EVENT	ACTION (Whether to manufacture part)	
	Manufacture part	Do not manufacture part
High demand (E_1)	100	70
Medium demand (E_2)	40	45
Low demand (E_3)	-20	10

The prior probabilities are: $P(E_1)$ = .35, $P(E_2)$ = .35, and
$P(E_3)$ = .30. A test market study of the potential demand for
the product is expected to report either a favorable (F) or
unfavorable (U) condition.   The relevant conditional
probabilities are: $P(F|E_1)$ = .60, $P(F|E_2)$ = .40, $P(F|E_3)$ = .10,
and $P(U|E_1)$ = .40, $P(U|E_2)$ = .60, $P(U|E_3)$ = .90.

17.4.10: The probability that the market research report will
be favorable; P(F) is:

(a)   .5526
(b)   .3684
(c)   .38
(d)   .62
(e)   .0789

Answer: (c)
Difficulty level: Medium

17.4.11: The probability that the event will be low demand, given that the market research report is favorable, is:

(a)   .5526
(b)   .3684
(c)   .38
(d)   .62
(e)   .0789

Answer: (e)
Difficulty level: Medium

17.4.12: The company's optimal decision strategy is:

(a)   if the market research report is favorable, then manufacture the part, if the market research report is not favorable, then do not manufacture the part
(b)   if the market research report is favorable, then do not manufacture the part, if the market research report is not favorable, then manufacture the part
(c)   do not manufacture the part
(d)   manufacture the part
(e)   none of these

Answer: (a)
Difficulty level: Medium

Use the table below to answer questions 17.4.13 - 17.4.15.

International Container, Inc., must decide where it will locate a new plant to produce a new packing product. The plant is to be constructed in London, Beijing, or New York $(A_1, A_2, \text{ and } A_3)$, and the major market for the product may be in Europe, the Far East, or the United States $(E_1, E_2, \text{ and } E_3)$. The profits (in millions of dollars) for the new product as presented in the payoff table below.

		ACTION (Plant Location)		
EVENT	Probability	London	Beijing	New York
Europe	.45	40	12	-5
Far East	.30	-10	50	18
United States	.25	28	-2	20

The past record of the marketing research company on similar studies has led to the following estimates of the relevant conditional probabilities:

Marketing Research Report

EVENT	Favorable (F)	Unfavorable (U)
Europe (E_1)	$P(F \mid E_1) = .60$	$P(U \mid E_1) = .40$
Far East (E_2)	$P(F \mid E_2) = .40$	$P(U \mid E_2) = .60$
United States (E_3)	$P(F \mid E_2) = .20$	$P(U \mid E_2) = .80$

17.4.13: The probability that the market research report will be favorable; $P(F)$ is:

(a) .44
(b) .56
(c) .6136
(d) .2727
(e) .3864

Answer: (a)
Difficulty level: Medium

17.4.14: The probability that the state of nature will be low demand, given that the market research report is favorable is:

(a) .44
(b) .56
(c) .6136
(d) .2727
(e) .3864

Answer: (d)
Difficulty level: Medium

17.4.15: The company's optimal decision strategy is:

(a) if the market research report is favorable, locate the plant in London, if the market research report is not favorable, locate the plant in new York
(b) if the market research report is favorable, locate the plant in New York, if the market research report is not favorable, locate the plant in London
(c) locate the plant in London
(d) locate the plant in Beijing
(e) locate the plant in New York

Answer: (c)
Difficulty level: Medium

Nonparametric Methods

18.1.1: Which of the following are advantages of using *nonparametric* methods?

I.   Generality
II. They are resistant to outliers and skewness
III. They do not require specific probability distributions

(a) I only
(b) II only
(c) III only
(d) I and II only
(e) I, II, and III

Answer: (e)
Difficulty level: Easy

18.1.2: We generally use nonparametric tests when the population is:

(a) Normal
(b) Binomial
(c) Nonnormal
(d) Poisson
(e) All of these

Answer: (c)
Difficulty level: Easy to Medium

Use the following data to answer Questions 18.1.3 - 18.1.5

A university professor would like to know if his teaching ability improves the quality of his students' scores on tests. He gave his ten students a test at the beginning of the semester and another at the end. He maintained that he could not assume the data follows a normal distribution. The results were:

Student	Test Score at Beginning	Test Score at End
1	20	80
2	49	79
3	72	60
4	59	99
5	75	75
6	95	45
7	86	36
8	10	50
9	1	81
10	52	22

18.1.3: The appropriate test for testing whether the two populations of student test scores are identical is the:

(a) Sign Test
(b) Mann-Whitney Test
(c) Wilcoxon Signed Rank Test
(d) Wilcoxon Signed Rank Test for Paired Data
(e) None of these

Answer: (c)
Difficulty level: Easy to Medium

18.1.4: If we are testing whether the populations of student scores on the two tests are identical at $\alpha = .05$, the test statistic and critical Z (or t) values are:

(a) $Z = .61$, $Z_{.05} = 1.645$
(b) $Z = .61$, $Z_{.025} = \pm 1.96$
(c) $t = .61$, $t_{.05,9} = 1.833$
(d) $t = .61$, $t_{.025,9} = \pm 2.262$
(e) $Z = -.61$, $Z_{.025} = \pm 1.96$

Answer: (b)
Difficulty level: Medium

18.1.5: If we are testing whether the populations of student scores on the two tests are identical at $\alpha = .05$, we should:

(a) reject the null hypothesis and conclude that differences between the two populations indicate that student scores were higher at the semester beginning
(b) reject the null hypothesis and conclude that differences between the two populations indicate that student scores were higher at the semester end
(c) not reject the null hypothesis and conclude that differences between the two populations indicate that student scores were higher at the semester end
(d) not reject the null hypothesis and conclude that there is not strong evidence of a significant difference between the two populations of student test scores
(e) reject the null hypothesis and conclude that there is no significant difference between the two populations of student test scores

Answer: (d)
Difficulty level: Medium

Use the following data to answer Questions 18.1.6 - 18.1.7

A manufacturing firm wants to know whether a difference in
task-completion times exists for two production methods. A
sample of 10 workers is taken, and each worker completes a
production task using each of the two production methods.
The production method that each worker used first is
selected randomly. Task completion times are in minutes.

Worker	Method 1 time	Method 2 time
1	11.2	10.5
2	10.2	9.8
3	11.6	11.1
4	10.9	11.3
5	11.2	10.3
6	11.6	11.5
7	11.0	11.0
8	12.2	11.6
9	10.7	11.2
10	11.6	10.8

18.1.6: If we are testing whether the populations of worker
task completion times for the two production methods are the
same at $\alpha = .05$, the test statistic and critical Z (or t)
values are:

(a) $Z = 2.04$, $Z_{.05} = 1.645$
(b) $Z = 2.04$, $Z_{.025} = \pm 1.96$
(c) $t = 2.04$, $t_{.05,9} = 1.833$
(d) $t = 2.04$, $t_{.025,9} = \pm 2.262$
(e) $Z = -2.04$, $Z_{.025} = \pm 1.96$

Answer: (b)
Difficulty level: Medium

18.1.7: If we are testing whether the populations of worker
task completion times for the two production methods are the
same at $\alpha = .05$, we should:

(a)   conclude that differences between the two populations
      indicate method 1 to be the better production method
(b)   conclude that differences between the two populations
      indicate method 2 to be the better production method
(c)   not reject the null hypothesis and conclude that the
      two population production methods are the same
(d)   maintain the status quo if we are currently using
      production method 1
(e)   none of these

Answer: (b)
Difficulty level: Medium

18.2.1: The sign test is:

(a)   a parametric test for the population mean
(b)   a parametric test for the population median
(c)   a nonparametric test for the population mean
(d)   a nonparametric test for the population median
(e)   a nonparametric counterpart to the t test for the
       equality of two means of two normal distributions

Answer: (d)
Difficulty level: Easy to Medium

Use the information below to answer Questions 18.2.2-18.2.3

In a study of consumer preferences to determine whether the
preferences for two brands of orange juice was equal, twelve
individuals were given unmarked samples of the two brands of
orange juice. The brand each individual tasted first was
randomly selected. After tasting the two products, the
individuals were asked to state a preference for one of the
two brands. Nine of the twelve individuals preferred brand
A. The remaining three preferred brand B.

18.2.2: If we are testing whether the percentage of people
who prefer orange juice brands A and B are equal at $\alpha = .05$
the appropriate null and alternative hypotheses are:

(a) $H_0: p = .50$; $H_a: p \neq 0$
(b) $H_0: p \geq .50$; $H_a: p < 0$
(c) $H_0: p \leq .50$; $H_a: p > 0$
(d) $H_0: \mu = .50$; $H_a: \mu \neq 0$
(e) H_0: Median $= .50$; H_a: Median $\neq .50$

Answer: (a)
Difficulty level: Easy

18.2.3: If we are testing whether the percentage of people
who prefer orange juice brands A and B are equal at $\alpha = .05$
we should:

(a)   conclude that the percentages of people who prefer the
       two brands of orange juice are the same
(b)   conclude that the percentage of people who prefer brand
       A of orange juice is significantly higher
(c)   conclude that the percentage of people who prefer brand
       B of orange juice is significantly higher
(d)   choose a sample size greater than 20 and redo the test
(e)   none of these

Answer: (b)
Difficulty level: Medium

Use the information below to answer Questions 18.2.4-18.2.5

A poll taken during a recent election asked 300 registered voters to rate the Democratic and Republican candidates in terms of best overall environmental policy. Results of the poll showed 160 rated the Democratic higher, 130 rated the Republican higher, and 10 rated the two candidates equally.

18.2.4: If we are testing whether the proportions of people who rate the Democratic and Republican candidates the best on environmental policy are equal at α = .05, the test statistic and critical Z (or t) values are:

(a) $Z = 0.20$, $Z_{.05} = 1.645$
(b) $Z = -0.20$, $Z_{.025} = \pm 1.96$
(c) $Z = 1.73$, $Z_{.05} = 1.645$
(d) $Z = -1.73$, $Z_{.025} = \pm 1.96$
(e)   None of these

Answer: (d)
Difficulty level: Easy to Medium

18.2.5: If we are testing whether the proportions of people who rate the Democratic and Republican candidates the best on environmental policy are equal at α = .05, our conclusion should be that:

(a)   there is not strong enough evidence to reject the null
      hypothesis that p = .50
(b)   there is strong enough evidence to reject the null
      hypothesis that p = .50
(c)   there is not strong enough evidence to reject the null
      hypothesis that p ≠ .50
(d)   there is strong enough evidence to reject the null
      hypothesis that p ≠ .50
(e)   none of these

Answer: (a)
Difficulty level: Medium

Use the information below to answer Questions 18.2.6-18.2.7

A few years ago it was reported that the median price for a home in a region of the eastern United States was equal to $95,000. A real estate agent wanting to test whether this is still true took a random sample of 50 homes and found that 31 of them exceeded $95,000 in price, 2 had prices of exactly $95,000 and 17 had prices lower than $95,000.

18.2.6: If we are testing whether the median home price in this region of the eastern United States equals $95,000 at α = .05, the test statistic and critical Z (or t) values are:

(a)   Z = -2.02, $Z_{.05}$ = 1.645
(b)   Z = 2.02, $Z_{.025}$ = ± 1.96
(c)   Z = -2.02, $Z_{.05}$ = 1.645
(d)   Z = 2.02, $Z_{.025}$ = ± 1.96
(e)   none of these

Answer: (d)
Difficulty level: Easy to Medium

18.2.7: If we are testing whether the median home price in this region of the eastern United States equals $95,000 at α = .05, the appropriate decision is to:

(a)   not reject the null hypothesis and conclude that the median home price in this region equals $95,000
(b)   reject the null hypothesis and conclude that the median home price in this region equals $95,000
(c)   not reject the null hypothesis and conclude that the median home price in this region is higher than $95,000
(d)   reject the null hypothesis and conclude that the median home price in this region is higher than $95,000
(e)   reject the null hypothesis and conclude that the median home price in this region is lower than $95,000

Answer: (d)
Difficulty level: Medium

Use the information below to answer Questions 18.2.8-18.2.9

A defense attorney wants to know whether there is a difference in divorce filings between husbands and wives. From courthouse records, it was found that in 100 divorce cases, the filing for divorce was initiated by the wife 62 times. Use α = .01.

18.2.8: The test statistic and critical Z (or t) values are:

(a)   Z = 2.40, $Z_{.05}$ = 2.575
(b)   Z = 2.40, $Z_{.025}$ = ± 2.33
(c)   Z = 0.48, $Z_{.05}$ = 2.33
(d)   Z = 0.48, $Z_{.025}$ = ± 2.575
(e)   None of these

Answer: (e)
Difficulty level: Easy to Medium

18.2.9: The appropriate conclusion for the test is to:

(a)   not reject the null hypothesis and conclude that the
      percentage of divorce filings is the same for men and
      women
(b)   reject the null hypothesis and conclude that the
      percentage of divorce filings is the same for men and
      women
(c)   not reject the null hypothesis and conclude that the
      percentage of divorce filings is significantly higher
      for women than men
(d)   reject the null hypothesis and conclude that the
      percentage of divorce filings is significantly higher
      for women than men
(e)   do nothing as results of the test are inconclusive

Answer: (d)
Difficulty level: Medium

18.2.10: The nationwide median hourly wage for a particular
labor group was found to be $10.00 per hour. A random
sample of 450 workers in a city in this labor group was
taken, with 230 workers having hourly wages that exceed
$10.00 per hour, 50 having hourly wages of exactly $10.00
per hour, and 170 workers having hourly wages less than
$10.00 per hour. If we test the hypothesis that the median
hourly wage in this city is the same as the nationwide
median hourly wage, the smallest level of significance (α)
that we can still reject the null hypothesis is:

(a)   .10
(b)   .05
(c)   .02
(d)   .01
(e)   for any α, we must accept the null hypothesis

Answer: (d)
Difficulty level: Medium

18.3.1: The Mann-Whitney test is:

(a)   a parametric test for the population mean
(b)   a parametric test for the population median
(c)   a nonparametric test for the population mean
(d)   a nonparametric test for the population median
(e)   a nonparametric counterpart to the t tests of the
      equality of two means for normal distributions

Answer: (e)
Difficulty level: Easy to Medium

18.3.2: Which of the following assumptions is(are) the Mann-Whitney test based upon?

I.   Both populations are normal
II. Both populations have identical variances
III. Random and independent sampling procedures

(a) I only
(b) II only
(c) III only
(d) I and II only
(e) I, II, and III

Answer: (c)
Difficulty level: Easy to Medium

18.3.3: When doing a test for whether two population means are equal ($\mu_1 = \mu_2$) and both populations are known to be nonnormal with unequal variances, the appropriate test is the:

(a) Sign Test
(b) Mann-Whitney Test
(c) Wilcoxon Signed Rank Test
(d) Wilcoxon Signed Rank Test for Paired Data
(e) None of these

Answer: (b)
Difficulty level: Medium

18.3.4: In an one-sided, upper tail Mann-Whitney test with *equal* sample sizes, the value of the test statistic is:

(a) the sum of the ranks for the smaller sample
(b) the sum of the ranks for the large sample
(c) the smaller rank sum
(d) the larger rank sum
(e) unable to be determined with the information given

Answer: (d)
Difficulty level: Medium

Use the following data to answer questions 18.4.1 - 18.4.2

Ten students were ranked for both their scholastic and
athletic ability, with the results summarized below.

Student	Scholastic Ability	Athletic Ability
A	1	5
B	2	6
C	3	9
D	4	3
E	5	8
F	6	4
G	7	10
H	8	1
I	9	7
J	10	2

18.4.1: To measure the degree and direction of the linear
association between the two sets of ranking, we should:

(a)   perform a Sign test
(b)   perform a Mann-Whitney test
(c)   perform a Wilcoxon signed rank test
(d)   compute the Spearman's rank correlation
(e)   none of these

Answer: (d)
Difficulty level: Easy to Medium

18.4.2: To measure the degree and direction of the linear
association between the two sets of ranking, the appropriate
decision to this problem (at $\alpha = .05$) is to:

(a)   not reject H_0 and conclude that the sample evidence
      does not support a strong correlation between
      scholastic and athletic ability
(b)   not reject H_0 and conclude that the sample evidence
      supports a strong and positive correlation between
      scholastic and athletic ability
(c)   to reject H_0 and conclude that there is a strong and
      negative correlation between scholastic and athletic
      ability
(d)   reject H_0 and conclude that there is no strong
      correlation between scholastic and athletic ability
(e)   reject H_0 and conclude that there is a strong and
      positive correlation between scholastic and athletic
      ability

Answer: (a)
Difficulty level: Medium

Use the following data to answer questions 18.4.3 - 18.4.4

The following data represent student rankings of four
students in two subjects, mathematics and english, with
these results:

	STUDENT			
SUBJECT	Abigail	Bill	Carla	Donny
Mathematics	3	2	4	1
English	2	3	1	4

18.4.3: To check whether there is a connection between
performance in the two subjects, the value of the
appropriate measure of the extent to which these two
rankings agree is:

(a)   1.0
(b)   0.5
(c)   0.0
(d)   -0.5
(e)   -1.0

Answer: (e)
Difficulty level: Easy to Medium

18.4.4: If we are testing whether there is a connection
between performance in the two subjects, based on rankings,
we should (at $\alpha = .10$):

(a)   not reject H_0 and conclude that there is a connection
(b)   reject H_0 and conclude that there is a connection
(c)   not reject H_0 and conclude that there is no connection
(d)   reject H_0 and conclude that there is no connection
(e)   none of these

Answer: (b)
Difficulty level: Medium

18.5.1: A nonparametric method for determining whether there
are differences between two populations based on two matched
samples where only one preference data is required is:

(a)   the Sign Test
(b)   the Mann-Whitney Test
(c)   Spearman's Rank Correlation
(d)   the Wilcoxon Signed-Rank Test
(e)   none of these

Answer: (b)
Difficulty level: Easy to Medium

18.5.2: A nonparametric test for the equivalence of two populations would be used instead of a parametric test for the equivalence of the population parameters if:

(a) no information about the samples is available
(b) no information about the populations is available
(c) the sample sizes are both very large
(d) the samples are dependent and not random
(e) the parametric test is always used to test the
     equivalence of two populations

Answer: (b)
Difficulty level: Easy to Medium

18.5.3: When using the ranking procedure for the following data: 10, 15, 15, 18, 20 the data values should have ranks of:

(a) 1, 2, 3, 4, 5
(b) 1, 2.5, 2.5, 4, 5
(c) 1, 2.5, 2.5, 3, 4
(d) 10, 15, 15, 18, 20
(e) 1, -2.5, 2.5, -4, 5

Answer: (b)
Difficulty level: Easy

Use the following data to answer questions 18.5.4 - 18.5.5

Students in statistics classes lasting two hours and fifteen minutes were asked whether they preferred a 15 minute break or to get out of class 15 minutes early, with no break. To test whether a difference in students' preferences exists, a random sample of 200 students was taken, with 70 preferring a 15 minute break, 30 having no preference, and 100 preferring to get out 15 minutes early.

18.5.4: The value of the test statistic (at α = .05) is:

(a) $Z_{.05}$ = 1.645
(b) $Z_{.025}$ = ± 1.96
(c) $t_{.025,2}$ = ± 4.303
(d) t = -2.30
(e) Z = -2.30

Answer: (e)
Difficulty level: Easy to Medium

18.5.5: At α = .05, we should conclude that:

(a) the same percentage of students prefer getting a 15 minute break to leaving 15 minutes early
(b) a significantly larger percentage of students prefer getting a 15 minute break to leaving 15 minutes early
(c) a significantly larger percentage of students prefer leaving 15 minutes early to getting a 15 minute break
(d) none of these
(e) we are unable to perform the test without additional information being given

Answer: (c)
Difficulty level: Medium

18.5.6: Independent random samples of faculty members at two colleges were taken, and annual salaries (in thousands of dollars) were recorded. The data are presented in the table below. At α = .01, is there a significant difference in the annual salaries of faculty members at the two colleges?

College 1	College 2
30.0	24.5
29.0	31.5
31.5	31.0
32.0	30.5
31.0	29.0
32.5	28.5
33.0	28.0

(a) Yes, since we cannot reject the null hypothesis
(b) Yes, since we can reject the null hypothesis
(c) No, since we can reject the null hypothesis
(d) No, since we cannot reject the null hypothesis
(e) It is uncertain, as the test results are inconclusive

Answer: (b)
Difficulty level: Medium

18.5.7: In a sample of 200 car owners, 78 indicated a preference for domestic cars, 102 indicated a preference for foreign cars, and 20 indicated no preference. If we want to test whether there is a significant difference in preference for domestic and foreign cars, we should (at α = .05):

(a)   not reject the null hypothesis and conclude that the proportions of people who prefer domestic and foreign cars are not significantly different
(b)   reject the null hypothesis and conclude that the proportions of people who prefer domestic and foreign cars are the same
(c)   not reject the null hypothesis and conclude that a higher proportion of people prefer foreign cars
(d)   reject the null hypothesis and conclude that a higher proportion of people prefer foreign cars
(e)   reject the null hypothesis and conclude that a higher proportion of people prefer domestic cars

Answer: (a)
Difficulty level: Medium

18.5.8: In a sample of 200 mall shoppers, 115 indicated a preference for fluoride toothpaste, 75 indicated a preference for non-fluoride toothpaste, and 10 indicated no preference. If we want to test whether there is a significant difference in preference for the two kinds of toothpaste, we should (at α = .01):

(a)   not reject the null hypothesis and conclude that the proportions of people who prefer fluoride and non-fluoride toothpastes are the same
(b)   reject the null hypothesis and conclude that the proportions of people who prefer fluoride and non-fluoride toothpastes are the same
(c)   not reject the null hypothesis and conclude that a higher proportion of people prefer fluoride toothpaste
(d)   reject the null hypothesis and conclude that a higher proportion of people prefer non-fluoride toothpaste
(e)   reject the null hypothesis and conclude that a higher proportion of people prefer fluoride toothpaste

Answer: (e)
Difficulty level: Medium

18.5.9: A small town has recently adopted a ban on smoking
in the towns restaurants and bars. A random sample of 100
residents who smoked prior to the ban is taken, with 55
approving of the ban on smoking, 30 not approving of the ban
on smoking, and 15 indicating no preference. A newspaper is
interested in knowing whether a majority of its citizens
support the ban on smoking. The appropriate null and
alternative hypotheses for the above test are:

(a) H_0: $p \leq .50$; H_a: $p > .50$
(b) H_0: $p \geq .50$; H_a: $p < .50$
(c) H_0: $p = .50$; H_a: $p \neq .50$
(d) H_0: $\mu_1 = \mu_2$; H_a: $\mu_1 \neq \mu_2$
(e) H_0: $\mu_1 \leq \mu_2$; H_a: $\mu_1 > \mu_2$

Answer: (c)
Difficulty level: Easy

18.5.10: A small town has recently adopted a ban on smoking
in the towns restaurants and bars. A random sample of 100
residents who smoked prior to the ban is taken, with 55
approving of the ban on smoking, 30 not approving of the ban
on smoking, and 15 indicating no preference. At $\alpha = .01$,
the appropriate decision is to:

(a) not reject the null hypothesis and conclude that the
     proportions of people who prefer and do not prefer the
     ban on smoking are the same
(b) reject the null hypothesis and conclude that the
     proportions of people who prefer do not prefer the ban
     on smoking are the same
(c) not reject the null hypothesis and conclude that a
     higher proportion of people in the town prefer the
     smoking ban
(d) reject the null hypothesis and conclude that a higher
     proportion of people in the town prefer the smoking ban
(e) reject the null hypothesis and conclude that a higher
     proportion of people in the town do not prefer the
     smoking ban

Answer: (d)
Difficulty level: Medium

18.5.11: If a null hypothesis that two populations are identical is rejected at α = .01 using a nonparametric test, then it is safe to assume that:

(a)   the two population means are not equal
(b)   the two population variances are not equal
(c)   neither the population means nor the variances of the two populations are equal
(d)   the population means and variances are equal
(e)   none of these

Answer: (c)
Difficulty level: Easy to Medium

18.5.12: A comprehensive examination in statistics is given to a small number of management and accounting majors to determine whether there is a significant difference in the knowledge of statistics among management and accounting students.   The following data summarize the scores of the 14 students.

Management	Accounting
72	81
76	84
80	79
78	86
83	91
79	88
71	86

If we want to test whether the populations of management and accounting students have the same median score using a nonparametric test, the appropriate conclusion to make based on the test described above (at α = .01) is:

(a)   the population median scores of management and accounting students are equal
(b)   the population mean scores of management and accounting students are equal
(c)   the population median scores of management students are significantly higher than those of accounting students
(d)   the population median scores of accounting students are significantly higher than those of management students
(e)   the two populations are identical

Answer: (d)
Difficulty level: Medium

Use the following data to answer questions 18.6.1 - 18.6.2

The management in a large assembly plant wishes to make changes in current assembly techniques, but union officials are afraid that the changes will result in lower wages for their workers, who are paid on the basis of output. A group of 14 employees, with varying levels of experience in the assembly process, was selected randomly. The assembly process was run using the old assembly techniques first and then using the new techniques. The hourly wages earned by the employees using the two techniques are presented below.

	TECHNIQUE			TECHNIQUE	
EMPLOYEE	New	Old	EMPLOYEE	New	Old
1	4.61	3.96	8	8.12	7.35
2	6.33	6.13	9	4.75	4.13
3	7.38	8.21	10	6.43	6.37
4	6.87	6.05	11	7.38	7.08
5	6.62	5.21	12	6.45	7.56
6	6.82	5.25	13	6.16	6.24
7	6.87	6.32	14	8.14	8.05

18.6.1: If we use a Wilcoxon signed-rank test to test the hypothesis that the median wage change is zero or less, the test statistic and critical Z (or t) values (at $\alpha = .10$) are:

(a) $Z = 1.73$, $Z_{.05} = \pm 1.645$
(b) $Z = 1.73$, $Z_{.10} = 1.28$
(c) $t = 1.73$, $t_{.10,13} = 1.350$
(d) $t = 1.73$, $t_{.05,12} = 1.782$
(e) $t = 1.73$, $t_{.05,13} = 1.771$

Answer: (b)
Difficulty level: Medium

18.6.2: If we use a Wilcoxon signed-rank test to test the hypothesis that the median wage change is zero or less, we should (at $\alpha = .10$):

(a)   not reject the null hypothesis and conclude that the median wage change is zero or less
(b)   reject the null hypothesis and conclude that the median wage change is zero or less
(c)   not reject the null hypothesis and conclude that median wages are higher under the new assembly technique
(d)   reject the null hypothesis and conclude that median wages are higher under the new assembly technique
(e)   none of these

Answer: (d)
Difficulty level: Medium

Index Numbers

19.1.1: Which of the following statements about index numbers are true?

I.   An index number measures the value of a time series in a period as a percentage of the time series in a base period.

II. The consumer price index (CPI) is correctly interpreted as a cost-of-living price index.

III. Most price indexes, such as the consumer price index, reflect changes made in the quality or availability of the products included in the index.

(a) I only
(b) II only
(c) III only
(d) I and III only
(e) None of these statements are true

Answer: (a)
Difficulty level: Easy to Medium

19.1.2: Index numbers are generally used:

(a) to show how the average value of a time series changes over time
(b) to express the variation of time series data values from some base value
(c) to express the variation of time series data values from the average of the data values
(d) to express the variation of time series data values from the variation in the data values
(e) to measure average price and unemployment levels

Answer: (b)
Difficulty level: Easy

19.2.1: A simple aggregate price index can be used
legitimately:

(a)   whenever all the prices are expressed in the same units
(b)   whenever equal amounts of each item are purchased
(c)   only when all the prices are expressed in the same
      units and equal amounts of each item are purchased
(d)   only when prices are expressed in different units or
      unequal amounts of different items are purchased
(e)   anytime, since it is the most often used index

Answer: (c)
Difficulty level: Easy to Medium

19.3.1: Given the years 1980, 1981, 1982, 1983 and related
data 23, 25, 27, 29, calculate the index with base year 1982
for 1980.

(a)   79.3
(b)   85.2
(c)   92.6
(d) 100.0
(e) 107.4

Answer: (b)
Difficulty level: Easy

19.3.2: Given the years 1980, 1981, 1982, 1983 and related
data 23, 25, 27, 29, calculate the index for 1982 using a
base year of 1980.

(a) 100.0
(b) 105.3
(c) 108.6
(d) 112.9
(e) 117.4

Answer: (e)
Difficulty level: Easy

19.3.3: The difference between the Laspeyres and Paasche price indexes:

(a)   lies in the choice of the base year
(b)   lies in the choice of the market basket
(c)   is that the Laspeyres index measures the change in cost for a fixed buying pattern whereas the Paasche index uses the nth-period quantities as weights for both the base period and the nth-period prices
(d)   is that the Paasche index measures the change in cost for a fixed buying pattern whereas the Laspeyres index uses the nth-period quantities as weights for both the base period and the nth-period prices
(e)   the Paasche index is used far more often than the Laspeyres index for measuring price changes

Answer: (c)
Difficulty level: Easy to Medium

19.3.4: Fisher's ideal index number is found by:

(a)   taking the square root of the Paasche index
(b)   taking the square root of the Laspeyres index
(c)   squaring the product of the Laspeyres and Paasche indexes
(d)   taking the square root of the product of the Laspeyres and Paasche indexes
(e)   none of these

Answer: (d)
Difficulty level: Easy to Medium

19.3.5: The index which measures the change in cost for a fixed buying pattern is called the:

(a)   Fisher index
(b)   Paasche index
(c)   Laspeyres index
(d)   weighted aggregate index
(e)   simple aggregate index

Answer: (c)
Difficulty level: Easy to Medium

19.3.6: State which of the following indexes does not use a base value of 100:

(a)   Laspeyres index
(b)   Paasche index
(c)   Physical-Volume index
(d)   Weighted aggregate index
(e)   All of these indexes use a base value of 100

Answer: (e)
Difficulty level: Easy to Medium

19.3.7: A value for the Laspeyres index of 363 means that:

(a)   The market basket you bought during the base year will now cost 263% more.
(b)   The market basket you buy will cost $363.
(c)   The market basket you bought during the base year will now costs 363% more.
(d)   There has been a sharp decrease in prices since the base year.
(e)   None of these.

Answer: (a)
Difficulty level: Easy

Use the information below to answer Questions 19.3.8-19.3.10

Tennis balls are sold in cans of three balls. The prices (in dollars) and quantities (in thousands of cans sold) for three major brands of tennis balls for the years 1988 and 1989 are presented in the table below.

| | PRICE | | QUANTITY | |
BRAND NAME	1988	1989	1988	1989
Penn	$2.25	$2.50	500	900
Wilson	$2.10	$2.40	750	1100
USTA	$2.50	$3.00	1000	1200

19.3.8: The value for the *Paasche* index to measure the increase in the aggregate price of tennis balls, using a base year of 1988 is:

(a)   115.75
(b)   116.35
(c)   86.40
(d)   85.95
(e)   116.05

Answer: (a)
Difficulty level: Easy to Medium

19.3.9: The value for the *Laspeyres* index to measure the increase in the aggregate price of tennis balls, using a base year of 1975 is:

(a)   115.75
(b)   116.35
(c)   86.40
(d)   85.95
(e)   116.05

Answer: (b)
Difficulty level: Easy to Medium

19.3.10: The value for Fisher's ideal index number to measure the increase in the aggregate price of tennis balls, using a base year of 1975 is:

(a)   115.75
(b)   116.35
(c)   86.40
(d)   85.95
(e)   116.05

Answer: (e)
Difficulty level: Easy to Medium

19.4.1: The period most often used currently by governmental agencies as a base rate is:

(a)   1928-1930
(b)   1968-1970
(c)   1978-1980
(d)   1982-1984
(e)   1980-1982

Answer: (d)
Difficulty level: Easy to Medium

19.4.2: In selecting a base year, one should consider:

(a)   the time criterion
(b)   comparability
(c)   whether the base year is a period of normal activity for the period of the series
(d)   the availability of data
(e)   all of these

Answer: (e)
Difficulty level: Easy

19.4.3: One may agglomerate several periods representing different stages in a business cycle to arrive at:

(a)   the time criterion
(b)   census of business cycles
(c)   fairly recent data
(d)   a normal base period
(e)   comparability

Answer: (d)
Difficulty level:

19.4.4: Index number series are used primarily for:

(a)   comparative purposes
(b)   determining base years
(c)   determining cost factors
(d)   determining physical-volume differences
(e)   economic data series like the GNP

Answer: (a)
Difficulty level: Easy to Medium

19.5.1: The most common use(s) of general purpose price indices is (are):

(a)   deflating
(b)   reducing to constant dollars
(c)   adjusting of dollar values
(d)   converting from nominal to real values
(e)   all of these

Answer: (e)
Difficulty level: Easy to Medium

19.5.2: An urban wage earner had an annual bonus of $6,780 in 1986. In 1992, the bonus rose to $7,460, but the worker was unhappy because he believed that his purchasing power had fallen. If the CPI in 1992 is 120 and the CPI in the base year of 1986 is 100, then the worker's purchasing power

(a)   rose by $680
(b)   fell by $680
(c)   rose by $563.33
(d)   fell by $563.33
(e)   stayed the same

Answer: (d)
Difficulty level: Easy to Medium

19.5.3: A firm's 1986 sales were $250,000 and in 1991 were $575,000. Given a wholesale price index (WPI) for the year 1991 of 125, the value of real 1991 sales in terms of 1986 base year dollars is:

(a) $ 4,600
(b) $ 200,000
(c) $ 312,500
(d) $ 460,000
(e) $ 718,750

Answer: (d)
Difficulty level: Easy to Medium

19.5.4: Which of the following is true about CPI?

(a)   It is often called the cost-of-living index.
(b)   It measures changes in prices of goods and services bought by urban residents.
(c)   Government fiscal and economic policies are influenced by its movement.
(d)   It has a great impact on the psychology of consumers.
(e)   All of these statements about the CPI are true.

Answer: (e)
Difficulty level: Easy

19.5.5: The salary of a certain construction worker in 1978 was $8,100. Ten years later, he was making $12,500. If the CPI in 1978 was 100 and became 182 in 1988, the worker's real salary in 1988 using a base year of 1978 is:

(a) $ 8,100
(b) $ 6,868
(c) $12,500
(d) $20,000
(e) $ 4,400

Answer: (b)
Difficulty level: Easy

19.5.6: The best-known index is probably the:

(a)   Paasche Price Index
(b)   Consumer Price Index
(c)   Gross National Product
(d)   Dow-Jones Industrial Average
(e)   Laspeyres Price Index

Answer: (b)
Difficulty level: Easy to Medium

19.5.7: A food products firm buys derivatives from two kinds of grains, wheat and barley, for use in its various products. The prices and quantities it purchased during two periods are shown below. Answer the questions below.

DERIVATIVE	PERIOD 1 Quantity	Price	PERIOD 2 Quantity	Price
Wheat	1,200 lbs	$1.40/lb	2,200 lbs	$2.50/lb
Barley	800 lbs	$2.30/lb	1,800 lbs	$4.20/lb

a.   Find the simple aggregate price chance index.
b.   Find the price change index which reflects the change in the total amount spent on these two derivatives between 1967 and last year.

Answer: a.   ($6.70/$3.70) x 100 = 181.1.   Note that the simple aggregate index in this problem makes more sense than in some problems since the quantity of measure is the same for both items--lbs. However, this only means that one pound of each item has gone up 81.1% since 1967 and does not reflect the usage levels.

          b.   This is the weighted aggregate index:
               (13,060/3520) x 100 = 371.02
Difficulty level: Easy to Medium

19.6.1: In 1991 the CPI was 135.5 and in 1992 it was 140.0. If a worker earned $12.54 per hour in 1991 and $13.00 per hour in 1992, and the worker worked 2000 hours in each year, then the worker experienced _____ in their purchasing power by a total amount (in annual dollars earned) of about _____.

(a) a rise, $4,764
(b) a rise, $7,259
(c) a fall, $4,764
(d) a fall, $7,259
(e) no change, zero

Answer: (a)
Difficulty level: Easy to Medium

Use the following table to answer questions 19.6.2 - 19.6.4

The data presented in the table below was taken from the May, 1991, issue of <u>Economic Indicators</u>. Included are annual values for nominal GNP (in billions of dollars), the implicit price deflator for GNP, and the consumer price index, for the years from 1986 to 1990.

Year	Nominal GNP (billions of $)	Price Deflator (1982 = 100)	Consumer Price Index (1982-1984 = 100)
1986	4231.6	113.8	109.6
1987	4515.6	117.4	113.6
1988	4873.7	121.3	118.3
1989	5200.8	126.3	124.0
1990	5465.1	131.5	130.7

19.6.2: Using the table above, the annual rate of inflation for the year 1987 to 1988 is:

(a) 3.32%
(b) 4.14%
(c) 7.93%
(d) 3.97%
(e) 4.46%

Answer: (b)
Difficulty level: Easy to Medium

19.6.3: Using the table above, the <u>percentage change in real GNP</u> from 1987 to 1988 (using 1982 constant dollars) is:

(a)   3.32%
(b)   4.14%
(c)   7.93%
(d)   3.97%
(e)   4.46%

Answer: (e)
Difficulty level: Medium

19.6.4: During the time period 1986-90, the smallest annual percentage change in the price deflator for GNP was during the year _____ and the largest percentage change in the annual inflation rate was during the year _____.

(a)   1986-87; 1986-87
(b)   1986-87; 1988-89
(c)   1986-87; 1989-90
(d)   1988-89; 1989-90
(e)   1989-90; 1989-90

Answer: (c)
Difficulty level: Easy to Medium

Use the following table to answer questions 19.6.5 - 19.6.9

The data presented in the table below was taken from the May, 1991, issue of <u>Economic Indicators</u>. Included are annual values of the consumer price index for the years 1987 - 1989 for each of the following sub-categories: food, shelter, and medical care.

Year	CPI, Food (1982-84 = 100)	CPI, Shelter (1982-84 = 100)	CPI, Medical Care (1982-1984 = 100)
1987	113.5	121.3	1_0.1
1988	118.2	127.1	138.6
1989	125.1	132.8	149.3

19.6.5: Using the table on the previous page, which of the three items experienced the *largest* percentage increase in the CPI during the period 1987-89?

(a)   food
(b)   shelter
(c)   medical care
(d)   food and shelter
(e)   shelter and medical care

Answer: (c)
Difficulty level: Easy to Medium

19.6.6: Using the table on the previous page, which of the three items experienced the *smallest* percentage increase in the CPI during the period 1987-89?

(a)   food
(b)   shelter
(c)   medical care
(d)   food and shelter
(e)   shelter and medical care

Answer: (b)
Difficulty level: Easy to Medium

19.6.7: If an individual spent the *same* percentage of income on food, shelter, and medical care in each of the years 1987, 1988, and 1989 (25% on food, 40% on shelter, and 35% on medical care), the value of the weighted average price index for goods purchased by the individual in 1987 is:

(a)   128.90
(b)   136.65
(c)   122.43
(d)   100.00
(e)   unable to be determined with the information given

Answer: (c)
Difficulty level: Medium

19.6.8: If an individual spent the *same* percentage of income
on food, shelter, and medical care in each of the years
1987, 1988, and 1989 (25% on food, 40% on shelter, and 35%
on medical care), the value of the weighted average price
index for goods purchased by the individual in 1988 is:

(a)   128.90
(b)   136.65
(c)   122.43
(d)   100.00
(e)   unable to be determined with the information given

Answer: (a)
Difficulty level: Medium

19.6.9: If an individual spent the *same* percentage of income
on food, shelter, and medical care in each of the years
1987, 1988, and 1989 (25% on food, 40% on shelter, and 35%
on medical care), the value of the weighted average price
index for goods purchased by the individual in 1989 is:

(a)   128.90
(b)   136.65
(c)   122.43
(d)   100.00
(e)   unable to be determined with the information given

Answer: (b)
Difficulty level: Medium

Use the table below to answer questions 19.6.10 - 19.6.12

The following items are included in a basic "market basket"
designed to measure a family's food consumption: milk,
bread, steak, and cheese. The prices and quantities for
each of these goods in 1985 and 1988 are presented in the
table below.

	1985		1988	
FOOD ITEM	Price	Quantity	Price	Quantity
Milk (quarts)	$0.30	150	$0.40	140
Bread (loaves)	$0.95	30	$1.15	40
Steak (pounds)	$2.00	15	$3.00	12
Cheese (pounds)	$1.80	10	$2.00	15

19.6.9: The value for the Paasche index to measure the increase in the aggregate price of the above market basket, using a base year of 1985 is:

(a)   129.75
(b)   128.24
(c)   131.28
(d)   138.27
(e)   121.76

Answer: (b)
Difficulty level: Easy to Medium

19.6.10: The value for the Laspeyres index to measure the increase in the aggregate price of the above market basket, using a base year of 1985 is:

(a)   129.75
(b)   128.24
(c)   131.28
(d)   138.27
(e)   121.76

Answer: (c)
Difficulty level: Easy to Medium

19.6.11: The value for Fisher's ideal index number to measure the increase in the aggregate price of the above market basket, using a base year of 1985 is:

(a)   129.75
(b)   128.24
(c)   131.28
(d)   138.27
(e)   121.76

Answer: (a)
Difficulty level: Easy to Medium